CODE NAME
MULBERRY

CODE NAME
MULBERRY

THE PLANNING - BUILDING & OPERATION
OF THE NORMANDY HARBOURS

GUY HARTCUP

Pen & Sword
MILITARY

To
Justin and Ellen
The story of a venture in which the imagination, invention and
engineering skill of the British were seen at their best.

First published in Great Britain in 1977 by
David & Charles (Publishers) Limited, Newton Abbott, Devon

Published in this format in 2006
and reprinted in 2011 & 2014 by
PEN & SWORD MILITARY
An imprint of
Pen & Sword Books Ltd
47 Church Street
Barnsley, South Yorkshire
S70 2AS

ISBN 978 1 84884 558 9

A CIP catalogue record for this book is
available from the British Library

Printed and bound in England by
CPI Group (UK) Ltd, Croydon, CR0 4YY

Pen & Sword Books Ltd incorporates the Imprints of Aviation, Atlas,
Family History, Fiction, Maritime, Military, Discovery, Politics, History,
Archaeology, Select, Wharncliffe Local History, Wharncliffe True Crime,
Military Classics, Wharncliffe Transport, Leo Cooper, The Praetorian Press,
Remember When, Seaforth Publishing and Frontline Publishing

For a complete list of Pen & Sword titles please contact
PEN & SWORD BOOKS LIMITED
47 Church Street, Barnsley, South Yorkshire, S70 2AS, England
E-mail: enquiries@pen-and-sword.co.uk
Website: www.pen-and-sword.co.uk

Contents

Glossary

ANCXF:	Allied Naval Commander Expeditionary Force
Baker Dolphin:	Extension to a floating pierhead
Baker Fender:	Crank-shaped concrete fender to protect a spud pontoon
Beetle:	Reinforced concrete float, or pontoon, to carry floating roadway
Bombardon:	Floating steel breakwater anchored by moorings
Bubble Break-water:	Device for suppressing waves, using compressed air
Buffer pontoon:	Ramp for beaching landing craft at a floating pierhead
CB:	Construction Battalion (Seabees); the shore and base construction force of the US Navy
CEC:	Civil Engineering Corps, US Navy
COHQ:	Combined Operations Headquarters
Corncob:	Assembly, fitting out and sailing of blockships for fixed breakwaters
COSSAC:	Chief of Staff to Supreme Allied Commander
CQR anchor:	Used by Landing Ships Tank. The anchor digs itself into the sea bed
Croc:	Tubular steel span for Hughes pier
D-Day:	Day of invasion; D-Day plus 1, one day after initial assault; D-Day plus 2, etc
DMWD:	Department of Miscellaneous Weapon Development in the Admiralty
DUKW:	Amphibious vehicles: D—year (1942); U—utility; K—front-wheel drive; W—two rear driving axles. In theory, they were able to carry ashore a $2\frac{1}{2}$-ton load in nets, but it was usually less
E-Boat:	Fast German motor torpedo boat
Erection tank:	Steel cylindrical tank used for connecting spans of floating roadway
Gooseberry:	Sheltered water formed by line of sunken ships

Grasshopper:	Floats equipped with gantries, used in erection of spans for the Tn5 project
Hippo:	Concrete caisson for Hughes pier
LCI (L):	Landing Craft Infantry (Large). Carried 250 troops
LCM:	Landing Craft Mechanised. Able to carry load of 30 tons
LCT:	Landing Craft Tank. According to size, able to carry loads from 30 to 300 tons
LSD:	Landing Ship Dock
LST:	Landing Ship Tank. Carried load of 2,150 tons when beaching; increased to 2,500 tons when berthing alongside floating pierhead. Held 50 vehicles or 20 vehicles and 22 tanks
Liberty ship:	Prefabricated cargo ship carrying over 7,000 tons
Lilo:	Collapsible floating breakwater made of canvas
Monolith:	Concrete caisson, later renamed Phoenix
MTL:	Motor Towing Launch (US Army)
Mulberry:	Artificial harbours for British and American beaches in Normandy landings
Neptune:	Plan for naval side of cross-Channel assault known as Operation Overlord
NL pontoons:	Naval Lighter pontoons (US Navy). Rectangular steel tanks assembled to form raft or jetty
NPL:	National Physical Laboratory
Overlord:	Code name for plan to liberate Europe agreed on at Quebec, August 1943
Phoenix:	Large concrete caisson made in different sizes and fitted with flooding valves to allow sinking at site to form breakwater
Pluto:	Pipeline Under the Ocean, for assuring the supply of fuel required by the mechanised forces
Port Construction Pontoon:	British equivalent of the NL pontoon. Standard size 15ft 6in long by 5ft 6in wide by 5ft 6in deep. Used in Far East
RAM/P:	Rear-Admiral Commanding Mulberry/Pluto— Rear-Admiral W. G. Tennant
RE:	Royal Engineers

Rhino Ferry:	Large oblong rafts formed by NL pontoons; propelled by outboard motors. Carried 30–40 vehicles
RASC:	Royal Army Service Corps
Shore Ramp Float:	Wedge-shaped float of light construction and shallow (6in at inshore ends) which connected floating bridge to shore
SLUG Boat:	Surf Landing Under Girder Boat used for mooring floats of floating roadway
ST:	Small tug. US Army harbour and river tug of 750hp
Swiss Roll:	Floating roadway of wire and planks
Tn 5:	Transportation 5 of the Directorate of Transportation, War Office
Whale:	Code name given to floating pierheads and roadways used in Mulberry to cross water and mud gaps and to berth coasters
WSA	War Shipping Administration. Operating agency of US merchant vessels carrying war cargoes
YMS	Small wooden minesweeper used by US Navy

Introduction

On an early spring day in 1944 a soldier looked out to sea from the skilfully-camouflaged concrete casemate of the coastal battery of 155mm guns on the Pointe de Hoc at the base of the Cherbourg peninsula. Below him the retreating tide revealed a variety of stakes and gate-like obstacles studding the gently shelving beach. They had recently been installed on the instructions of the commander of the German Army Group B, Field-Marshal Erwin Rommel. Beyond the beach, among the low bluffs covered with undergrowth, were areas ringed by barbed wire. Within them the soldier knew that hundreds of mines were buried. Twenty miles inland units of a panzer division were on an exercise; their vehicles were tucked into the tall hedgerows and troops in field-grey clattered about the squares of the grey stone Norman villages. Apart from this military activity and the occasional passage of hostile bomber formations overhead on their way to Brest or another Atlantic port, the possibility of a battle in the quiet countryside seemed remote. Cattle grazed in the small green fields and farmers were doing their spring sowing, walking behind their stalwart farm horses.

Across the Channel in London, dingy and battered after four years of war, the barrage balloons swayed listlessly at their moorings. In a smoke-filled room behind the plain brick facade of Norfolk House in St James's Square, in the heart of the West End, and now the headquarters of the Chief of Staff to the Supreme Allied Commander (COSSAC), a number of aerial photographs of the scene described above were being closely examined by a group of army and naval officers. They were members of the planning staff for the liberation of Europe. They knew what the soldier on the Normandy cliff could only speculate on, the location and approximate timing of the long-anticipated Allied cross-Channel assault.

For months past the pros and cons of possible landing beaches had been endlessly discussed. Three immutable factors dictated the choice of the planners. First and foremost, surprise had to be achieved; secondly, the beaches had to be within

11

range of fighter cover from the airfields of southern England so that the maximum effect of the hard-won Allied air superiority could be obtained; thirdly, ports had to be available within a reasonably short time after the assault to ensure a flow of supplies to the armies as they advanced through France to Germany.

Several plans for the invasion had been worked out in detail and then shelved. Allied strategy demanded that operations in the Mediterranean theatre should take precedence from the latter part of 1942 and for the whole of 1943. This did not mean that preparations for a cross-Channel assault came to a stop. Far from it. In May 1943 the Combined Chiefs of Staff agreed that a full-scale attack should be launched not later than May 1944. Planning then went ahead for a three-divisional assault supported by airborne troops and commandos. Six divisions would back up the assault and a further twenty divisions would reinforce the beachhead.

Amphibious operations are invariably complicated. The Allies had acquired considerable experience in techniques in the landings in North Africa, Sicily and Italy but the so-called 'Festung Europa' was different. All the likely beaches were covered by fire; a system of self-sufficient defences stretched along the northern coastline of France into Belgium and Holland. If the landings were to be successful the outer defences would have to be breached and a strong beachhead established.

The Allied Staffs knew that the enemy's strategy had two aims. The first—and it was abundantly clear from the obstacles on the beaches—was to prevent the invaders from gaining a foothold. By rapidly deploying his mobile reserves, Rommel would throw them back into the sea. If the Allies were going to win this first, crucial round they would have to build up their forces in the narrow beachhead at the same rate as, or more rapidly than Rommel could bring up his reinforcements. The second aim of the enemy appeared to be to prevent the Allies from securing ports, Cherbourg, in particular, through which a flow of men and supplies could be maintained.

At Norfolk House, the planners had calculated how many tons of supplies would be required for the first three weeks of operations. They based their figures on a three-divisional assault, building up to a strength of ten divisions by D-Day plus 5 followed by the subsequent landing of a division each day. At this rate some 10,000 tons were required on D-Day plus 3, 15,000 tons on D-Day plus 12 and 18,000 tons on D-Day plus 18. As will be explained later, these figures had to be increased by two more divisions when the assault was strengthened by General Montgomery on his appointment as tactical commander of the Allied land forces.

How were the vital ammunition, petrol and rations going to be discharged? Provided the ports in Normandy and Brittany could be captured more or less intact, and in reasonable time, they could support about thirty divisions. But frontal assaults on ports were too costly, and from experience in the Mediterranean it was anticipated that the Germans would demolish the port facilities, making them unusable for some time.

There were four or five minor ports, mainly used by fishing vessels, in the area chosen for the assault. But they could not be expected to discharge more than 1,300 tons per day in the first two weeks. Cherbourg itself, originally designed for passengers disembarking and embarking on transatlantic liners rather than for discharging cargo, was unable to maintain the invasion forces; its capacity was only 3,750 tons after one month. The Brittany ports were unlikely to be available until after some two months of operations.

Reluctantly, the planners concluded that the initial build-up would have to take place over open beaches. Some eighteen divisions would have to be supported in this way during the first month of operations and twelve in the second. By the end of the third month, with any luck, the ports would have been sufficiently restored to working order to make discharge over the beaches no longer necessary.

All these estimates were, of course, theoretical. What would happen if continuous bad weather prevented ships from discharging? Careful studies were made of the weather along that part of the French coast over the previous ten years and these showed that June was the most favourable month for beach-landing craft. But, according to the records, fair spells of weather of only four days in succession could be expected in every month between May and September. This meant that from D-Day plus 4 it was likely that the discharge of supplies would be interrupted and to allow for this, the daily discharge would have to be increased by around 30 per cent. Allowance also had to be made for overcoming the difficulties of moving and distributing a mass of supplies in a narrow, congested beachhead.

Moreover, the navy warned that landing craft continuously grounding on the beaches might damage their bottoms, thus reducing the available lift and putting the whole operation in hazard. For this reason berthing facilities were considered to be especially important. In fact, Admiral Sir Bertram Ramsay, Commander of the Allied Naval Expeditionary Force, said emphatically that he would not willingly embark on the operation without them.

It must now be apparent—and the planners were in agreement over this—firstly, that unless sheltered water was provided by artificial means the operation would be at the mercy of the weather; and the English Channel is notoriously fickle at any time of the year. Secondly, special berthing facilities were needed within the sheltered water, especially for the discharge of vehicles. In short, an artificial harbour was required. More than that, it had to be a harbour whose components could be built in safety in friendly territory, then towed across the Channel and installed off the landing beaches. But first it had to be designed, tested and built, and the troops trained to operate it. The reader does not need to be an engineer to appreciate that to develop and build such equipment required great skill. Still more significant, time was short.

So it was that on the spring day in question, an observer on the promenade at Rye in the Isle of Wight could see strange-looking pontoons with chimney-like structures rising from them being towed, and sections of what appeared to be floating bridges grinding and creaking at their anchorages. Stranger still, further down the coast at Selsey Bill, he would have seen great concrete caissons rising out of the water.

What follows is the story of how all this equipment was designed, assembled, towed across the Channel—a distance of 100 miles—and installed on the far shore in the wake of the Allied landings in June 1944.

1 Concept and Command

The two artificial harbours which were to be installed, for the Normandy landings, one in the British and one in the American sector, were given the code name Mulberry. The choice of this name was entirely fortuitous. It covered the entire operation of the harbour; other code names were given to the various components.

The purpose of the Mulberries was to provide shelter for shipping and, secondly, facilities for discharging cargo. The shelter consisted, firstly, of concrete caissons (Phoenixes) sunk in line offshore and, secondly, floating steel tanks of a cruciform shape (Bombardons) moored about three miles out to sea. The unloading facilities (Whale) comprised floating pierheads, capable of adjusting to the rise and fall of the tide and connected to the shore by roadways, susceptible to the motion of the sea, laid on floats which were secured to moorings attached to anchors in the seabed.

Several months before D-Day it was decided to widen the area of the assault. Shelter for the additional landing craft was provided by crescent-shaped breakwaters (Gooseberries). These were formed by scuttling unserviceable merchant ships (Corncobs) which sailed across the Channel under their own steam. Block-ships were also introduced into the Mulberries to provide immediate protection for assault and other small craft and to enlarge the area of sheltered water.

The necessity for landing an army in enemy-occupied territory had not been foreseen by the naval and military planners before Dunkirk. When preparations for liberating the continent began, the problem arose of how to maintain a large force on the other side of the Channel. In North Africa and Italy the ports had been undefended and restored to working order within a relatively short time. But in north-west Europe the disastrous raid on Dieppe in August 1942, in which over 900 officers and men were killed and some 2,000 captured, had shown that the Germans were determined to deny the use of ports to the Allies for as long as possible.

Even earlier, however, the strategic situation had driven the

15

Prime Minister, Winston Churchill, to promote plans for piers and pierheads to be moored off open beaches. The intermediary for Churchill's demand was Admiral Lord Louis Mountbatten, the energetic and resourceful Chief of Combined Operations. He had under him a small inter-service staff who were responsible for developing techniques of amphibious warfare. But they had not the necessary capability for developing the quantity of floating equipment which would be required.

The War Office was responsible for port construction and operation; the Admiralty was responsible for the general operation of shipping within the harbour. Following the established principle that ports are built and maintained by civil engineers, leaving the seamen to take care of their ships, the army was given responsibility for the construction and organisation of harbours while the navy took charge of naval docks and bases.

Early in 1941 Maj-Gen D. J. (later Sir Donald) McMullen, Director of Transportation in the War Office, formed a new branch of his directorate called Transportation 5 (Tn5) to be responsible for port engineering. He put Major (later Brigadier Sir Bruce) White in charge of it. Bruce White had an all-round knowledge of engineering and was a member of the civil, mechanical and electrical engineering institutions. He had served in the Royal Engineers in World War I and since then had specialised in the development of transport systems at home and abroad.

Bruce White must be given credit for appreciating, firstly, that in operations against enemy-occupied territory, it would be necessary not only to repair devastated ports but actually to *build* them; secondly, that for this purpose there would have to be a register of civil engineers, particularly those who had had experience of port engineering in various parts of the Empire and who were now either misemployed in one of the services or ministries, or outside them.[1]

Tn5 therefore became responsible for forming and controlling the port construction and repair and port maintenance companies, and for the allocation and control of port repair ships (tramp steamers of about 800 tons fitted with equipment for clearing debris from harbour beds, pumps, compressors, fire fighting appliances and facilities for repairing machinery), dredgers, floating cranes and inland water transport. Other branches of the directorate dealt with port layout and the operation and construction of railways. As the war spread, its responsibilities became world-wide. From the register made by Bruce White some 150 engineers joined Tn5 or were posted to units in the field. Being civilians in uniform, they were sometimes looked on with suspicion by regular engineer officers and at times caused confusion, as when, for example, the captain of a port repair ship assumed the rank of major when ashore. It is worth noting that the Americans copied some of Tn5's ideas, in particular port repair ships.

Tn5's first big task was to construct Nos 1 and 2 Military Ports on Gareloch off the Clyde (now a nuclear submarine base) and at Cairn Ryan in Wigtownshire.[2] The intention was that they should provide deep-water berths for shipping engaged in military movement, thereby relieving civil ports on the English west coast, heavily congested after it became impossible to make full use of facilities on the south-east and south coasts. They were also used by American forces entering Britain. Their construction and dredging was entirely carried out by sappers, a number of whom benefited from the experience when the time came to install Mulberry.

The fact that in the early stages the soldiers were responsible for the planning and execution of Mulberry became a source of vexation for the Admiralty which, of course, was inextricably concerned with harbour layout and cross-Channel movement. But, unlike the civil engineering side of the US Navy, the department under the Civil Engineer-in-Chief, A. F. (later Sir Arthur) Whitaker, was not comparable with Tn5 in that his engineers, though knowledgeable in port work, had had no experience of building them under fire. This inevitably led to arguments relating to design and priorities. Bruce White has always maintained, with understandable vehemence, that the double responsibility that was later, as will be seen, imposed because the navy never formed a port construction force, was a severe handicap to the execution of such a complex operation. Certainly, in retrospect, it seems strange that at such a late stage in the war the benefit of a combined headquarters was not applied to the direction of the Mulberries.

However, the Admiralty did make one positive contribution. This was through a temporary wartime creation called the Department of Miscellaneous Weapons Development (DMWD). Formed originally at the outbreak of war to investigate aspects of anti-aircraft gunnery, it was later expanded to cover the whole field of naval warfare, including proposals for amphibious operations. An imaginative Canadian scientist, but also an RNVR officer, Charles F. (later Sir Charles) Goodeve gave DMWD its character. He enlisted the services of a number of young physicists, chemists, engineers and others from a variety of un-nautical like professions, equally unorthodox and irreverent in their attitude to authority.[3] The Department of Naval Construction also played a part in the building of the floating breakwaters.

Makers of Allied strategy in 1943 concluded that, in spite of pressure from the Russians (reflected in a popular demand for a 'second front') and mainly for logistical reasons, a full-scale assault could not be launched across the Channel before May 1944. In the meantime, the main Allied pressure was to be in the Mediterranean, an entry into Europe being made via Sicily and Italy.

By the end of July 1943, however, plans for landings in Normandy (code word 'Overlord') were far advanced. The planners appreciated that during the initial stage of the assault supplies would have to be brought in over open beaches in the absence of a port and that an artificial harbour would be needed. It would have to be towed over from England as neither facilities nor time were available on the far side.

The Overlord plan was examined and accepted by the Prime Minister, President and the Combined Chiefs of Staff at Quebec that August. Until then, the tentative proposals for artificial harbours were known only to the planners and to a few officers at Headquarters Combined Operations. In order to bring home their function to the British Chiefs of Staff, Lord Mountbatten arranged a lecture on the *Queen Mary* taking the Prime Minister and his distinguished party to North America.[4] A selection of senior officers was assembled in one of the ship's luxurious bathrooms where a demonstration was given by Professor J. D. Bernal, the eminent physicist and one of Mount-

I Naval Lighter Pontoon with inboard engine. The rectangular tanks of this versatile US invention could be used for making Rhino ferries, causeways or as pontoons for carrying a floating roadway (*Public Record Office*)

batten's scientific advisers. Standing on a lavatory seat, Admiral Sir Dudley Pound, First Sea Lord, invited his colleagues to imagine the shallow end of the bath as a beachhead. Bernal now floated, with the assistance of Lt-Cdr D. A. Grant, a fleet of twenty ships made of newspaper. Grant was then requested to make waves with the aid of a back brush. The fleet sank. A Mae West lifebelt was then inflated and floated in the bath so as to represent a harbour. The fleet of paper boats was then placed inside it. Once more Grant applied his brush vigorously to create waves, but this time they failed to sink the fleet. By such simple means were the senior officers convinced of the importance of sheltered water.

Equally important, of course, was the requirement for facilities for discharging cargo. (Hitherto the combined operations planners had not envisaged anything more ambitious than the beaching, or 'drying out' of landing craft.) Having discharged their load, they then had to wait until refloated by the next high tide. The Civil Engineer Corps of the US Navy was then experimenting at Rhode Island with rectangular steel pontoons (called Naval Lighter pontoons invented by Capt John N. Laycock, USN). (Plate I.) They were filled with air and could be strung together to form rafts (Rhino ferries) to carry tracked or wheeled vehicles from ship to shore.[5] They could also be flooded so as to form causeways which could be rapidly constructed on a captured beach and alongside which small craft could discharge or troops be brought ashore dryshod. This equipment was to prove of the utmost value in the Mulberries but it did not match the elaborate pierhead and floating roadway recently developed by the British. The latter also developed similar pontoons (Port Construction Pontoons); a few of them were used for lighterage in Normandy but they really came into their own in Burma.

The Americans, unlike the British, brought a strong team of naval and military engineers to Quebec. But when Lord Mountbatten in an important paper defined for the first time the implications involved in the erection of an artificial harbour, it was clear that more evidence on the British proposals for floating pierheads and caissons was required.[6] An urgent signal was sent to London summoning the Tn5 engineers and an Admiralty team which had been developing floating break-waters. They came out post-haste by air.

The Combined Chiefs of Staff, after seeing the films and working models brought over by Tn5, approved the proposal for British and American artificial harbours which were, in due course, known collectively as Mulberry.

From Quebec the British and American engineers went on to Washington where in a short, but intensive, period of collaboration the basic design of the harbours was worked out. Initial experiments with concrete breakwaters also took place. The requirements were that the American harbour (Mulberry A) should be capable of unloading 5,000 tons and the British (Mulberry B) 7,000 tons per day. The pierheads should be capable of berthing up to eleven coasters, according to their size. The pierheads were to be protected either by floating or

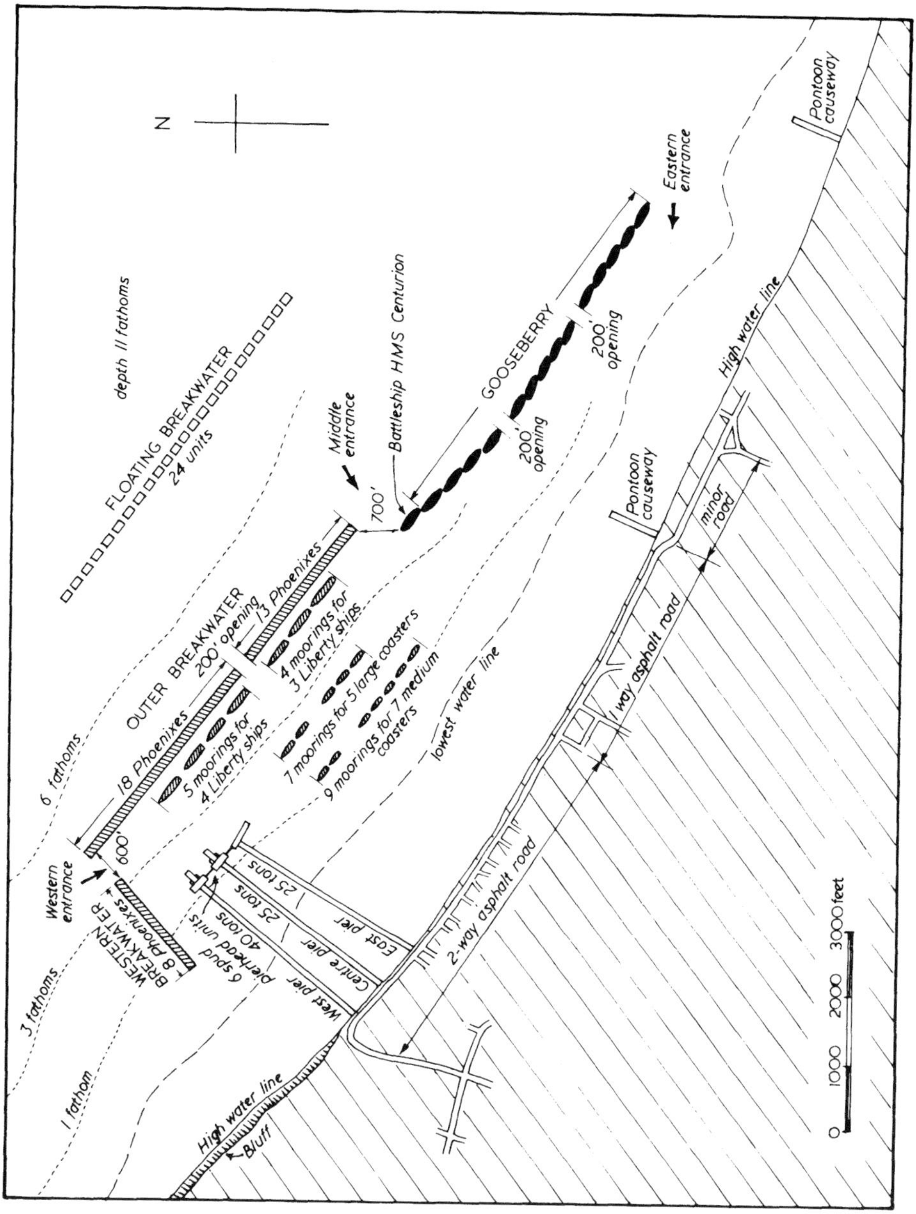
N
FLOATING BREAKWATER
24 units
depth 11 fathoms
6 fathoms
3 fathoms
1 fathom
OUTER BREAKWATER
18 Phoenixes
200' opening
13 Phoenixes
5 moorings for
4 Liberty ships
4 moorings for
3 Liberty ships
7 moorings for 5 large coasters
9 moorings for 7 medium
coasters
lowest water line
Middle
entrance
700'
Battleship HMS Centurion
GOOSEBERRY
200'
opening
200'
opening
Eastern
entrance
Western
entrance
600'
WESTERN
BREAKWATER
8 Phoenixes
6 spud
pierhead units
40 tons
25 tons
25 tons
West pier
Centre pier
East pier
High water line
Bluff
2-way asphalt road
1 way asphalt road
minor
road
Pontoon
causeway
High water line
Pontoon
causeway
0
1000
2000
3000 feet

Fig 1 Plan of Mulberry A

sunken breakwaters, or both. They had to be capable of being towed across the Channel and they had to be large enough to accommodate up to ten Liberty ships anchored within their shelter. In addition to the shipping unloading at pierheads, the sheltered water would enable DUKWs, Rhino ferries, lighters and barges to carry cargo from ship to shore. The harbours were to be fully in operation a fortnight after D-Day.

These proposals were immediately approved by the Combined Chiefs of Staff (see Figs 1 and 2). On 3 September 1943 they defined responsibilities on a national basis for building the components.[7] Bearing in mind the hazards and difficulties of crossing the Atlantic, it was clear that the work would have to be done in the United Kingdom. The British were therefore to begin work at once on the design and construction of concrete caissons, of which about $2\frac{1}{4}$ miles would be required, and were to continue with the development of floating breakwaters. They were also to press ahead with the development and production of fifteen floating pierheads and six miles of floating roadways. The Americans were to experiment with the alternative possibility of a floating breakwater made up of ships anchored in line. Both nations were to comb out all available tugs, particularly those of the ocean-going class. For such a great armada of units without any propulsive power of their own, tugs were a vital requirement. At least 130 were needed.

The British were now faced with the immense problem, in a country which had already been at war for three years and whose productive capability had been strained to the utmost, of building components for two harbours, each enclosing roughly two square miles of water, ie, a space bigger than that of Dover harbour. Moreover, the task had to be completed within seven months. Dover harbour had taken seven years to build. How was the work to be allocated and the responsibilities defined? On these issues the project nearly came to grief.

In fact, the first step had already been taken. The British Chiefs of Staff had appointed Sir Harold Wernher as Co-ordinator of Ministry and Service Facilities. Sir Harold was a wealthy and energetic businessman and the first chairman of the British Electrolux Co. He had married into the Russian royal family and his brother-in-law was Lord Mountbatten.

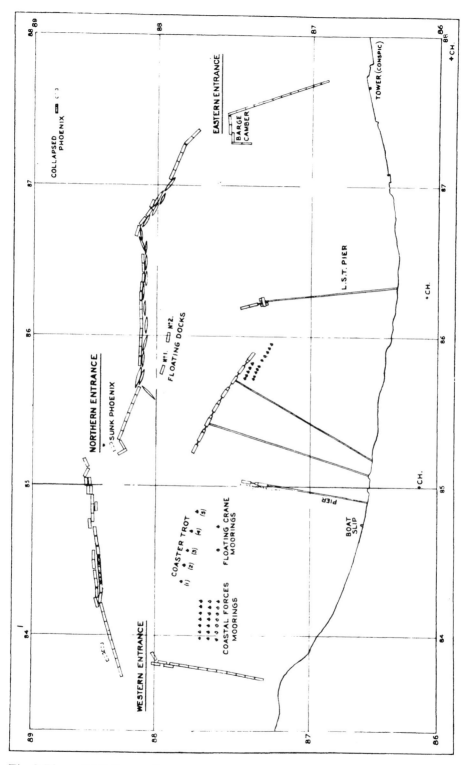

Fig 2 Plan of Mulberry B excluding floating breakwater

As a Territorial officer, he had already been serving in Combined Operations Headquarters. He had been responsible for overseeing the preparation of the south coast for a cross-Channel assault. Now, transferred to the staff of the Supreme Commander, Wernher had to act as liaison between the Supreme Commander, the Admiralty, the War Office, the appropriate US authorities and the British departments concerned with the development of ports, sheltered anchorages and landing facilities and to ensure that the whole Mulberry programme requirements were met and delivered in time.[8] It was a heavy responsibility which Wernher, able though he was, had not the authority to execute properly.

While there was no difficulty about allocating the responsibilities of the Ministries of Works, Production, War Transport and Supply, those of the Admiralty and War Office proved to be much more contentious. Before Quebec the Chiefs of Staff had decided that the Admiralty should broadly be responsible for sheltered anchorages, floating equipment and blockships. The War Office was to be responsible for the development of all other port and unloading equipment. But after Quebec, due to the importance attached to the provision of concrete breakwaters, the War Office became responsible for the design of the caissons. It also continued to supervise the development of pierheads and piers. The Admiralty was going ahead with floating breakwaters meanwhile.

Following the ruling stated earlier and also because of the lack of any response from the Admiralty, Tn5 had taken responsibility for the layout of the harbours. It was this aspect that the Admiralty was unable to stomach. For security reasons, ostensibly, Tn5 had been reluctant to reveal to the Admiralty what plans its port engineers were preparing. At the same time the Admiralty was becoming increasingly worried about the adequacy of berthing facilities, the setting up of navigational aids, the time it would take to complete the installation of the harbours off the French coast, and other technical matters.

Largely as a result of agitation by Capt Harold Hickling, RN, a brash and outspoken New Zealander and naval liaison officer to Wernher (he was later appointed naval officer in charge of Mulberry B), the Admiralty became convinced that the War Office was treating Mulberry on too parochial a level, and by refusing to divulge their plans were keeping the project, as it were, 'inside an oyster'.[9] As Wernher had inadequate authority to co-ordinate the two service departments, on 10 December the Vice-Chiefs of Staff were forced to intervene. Lt-Gen Sir Archibald Nye, Vice-Chief of the General Staff, said to Wernher: 'You are a co-ordinator. Why don't you co-ordinate?' The latter simply replied: 'You cannot co-ordinate people who refuse to be co-ordinated.'[10]

The issues were finally resolved on 15 December at a meeting chaired by General Morgan, Chief of Staff to the Supreme Allied Commander. A phalanx of admirals faced the army supply and transport officers, the most senior of whom was a major-general. Bruce White later described the ensuing discussion as 'one of the great battles of the war'. The upshot was that the Admiralty became responsible for towing all the components across the Channel and for the layout and siting of the harbours. They were also to be responsible for surveying the harbour sites, siting breakwaters, and for marking navigational channels and moorings. The War Office was to remain responsible for the design and building of the caissons, for their protection from air attack and for parking them until required for towing to the other side of the Channel. The army was also responsible for sinking them in position. It also continued to be responsible for construction and installation of all discharge facilities.[11]

When construction of the components was completed, Wernher would hand over responsibility to Rear-Admiral W. G. (later Sir William) Tennant, who held the title Rear-Admiral Mulberry/Pluto.* Tennant was to direct the cross-Channel towing operation and generally supervise the working of the two ports.

The installation of Mulberry B itself came under a joint naval–military command, namely that of Capt C. H. Petrie, RN, and Brigadier A. E. M. Walter, RE. Petrie was a hydrographer and Walter had first shown his ability as a staff officer

---

*Pipeline Under the Ocean: it assured the uninterrupted supply of fuel upon which the mechanised forces depended for mobility.

by collaborating with the navy during the evacuation from Dunkirk, later serving as a staff officer in Washington and South-east Asia.

Although the British were responsible for the construction of the Mulberry components, the Americans undertook to install their own harbour. For this purpose they were able to draw upon the services of their naval Corps of Civil Engineers. Mulberry A, therefore, unlike the British harbour, was exclusively a naval responsibility. Installation was to be carried out by Force 128, more commonly known as Force Mulberry, under Capt A. Dayton Clark, USN—a hard taskmaster and disciplinarian, who had had considerable experience of dealing with the Royal Navy.

The personnel for Force 128 were provided by No 108 Construction Battalion (CB) from No 25 Naval Construction Regiment.[12] The CBs (popularly known as 'Seabees') were a wartime formation composed of carpenters, electricians, welders and wharf builders, and were employed in building training stations and supply depôts, and assembling pontoon units for Rhino ferries. No 108 Construction Battalion provided crews for the caissons and Whale tows during the passage across the Channel and was responsible for installing, operating and maintaining the equipment on the far shore. The remaining three battalions manned the Rhino ferries and pontoon causeways. The CBs were much in demand in the Pacific theatre of war and had only been obtained after a wrestle with the Pentagon in Washington.

American collaboration with the British depended on a somewhat tenuous system of liaison. But as months and then weeks went by to D-Day, they became increasingly uneasy about the prospect of the components being completed in time. The British were being over-complacent. The truth was, and it was not unappreciated by the British, that the odds *were* against completion by D-Day. There were four factors which militated towards this gloomy conclusion. First, at this late stage of the war there was an acute shortage of labour, particularly welders, steel fixers, riveters, electricians, scaffolders and carpenters. Secondly, there was an equally acute shortage of the materials, such as steel, which were required for so many items of equipment. Thirdly, because of the need for shipbuilding and repair, there was a shortage of dry docks for the building of items like the caissons and pierheads. Fourthly, there was an inadequate number of tugs for the jobs that had to be done, not only before, but for some weeks after, D-Day. It is against all these difficulties, problems of command, co-ordination, shortage of manpower and materials and, above all, lack of time, that the development of the harbours must be considered.

2 Piers and Pierheads

What seemed to be the intractable problem impeding an assault in the Baie de Seine area was the very shallow beaches stretching far out into the Channel at low tide. A 2,000-ton coaster, or larger stores ship, could not approach nearer than about a mile from the shore and at high tide the gap between ship and shore was about 60ft. The difference between the rise and fall of tide was as much as 20ft during spring tides. The possibility of beaching coasters and landing craft has already been mentioned, but it was impossible to ensure a continuous flow of supplies by this means alone. In the early, critical stage of extending a beachhead, continuity of ammunition supplies, for example, could easily make the difference between success and failure.

There seems to be no doubt that it was Churchill who stimulated the development of special equipment to overcome this problem. In a memorandum addressed to Mountbatten on 30 May 1942 he enumerated the advantages of a floating pier as opposed to a trestle structure built out from the shore which was the standard military solution. One of the principal drawbacks of the latter was the long time it would take to build. But the most original part of the Prime Minister's note was the concluding paragraph couched in characteristic terms.[1] The pierheads, Churchill declared, '*must* float up and down with the tide. The anchor problem must be mastered. The ships must have a side flap out in them and a drawbridge long enough to over-reach the moorings of the piers. Don't argue the matter. The difficulties will argue for themselves.'

The development of equipment for the discharge of supplies was entrusted to Tn5. In charge was Major Vassal Steer-Webster, an unusual and colourful character such as projects like Mulberry tended to attract. Steer-Webster had served, and been wounded in, World War I. Afterwards he claimed to have been a technical adviser in India and the Middle East. In fact, Steer-Webster's technical knowledge was outmatched by a flair for showmanship which attracted the attention of very senior officers and even gained him the ear of the Prime

Minister. At the same time it was inevitable that he should irritate the professional engineers who only wanted to get on with the job. Nevertheless, he played a useful rôle in promoting Whale equipment when it had to compete with less worthy proposals.

Early in 1943 a special unit was formed by Tn5 and given the cumbersome but all-embracing title of No 1 Transportation Fixed and Floating Equipment, Development and Training Depot RE. It became responsible for the development and training in the installation of the piers which were used in Mulberry. In command was Major (later Lt-Col) J. G. Carline, a former engineer on the Assam railways, who had been drawn into the Tn5 net. He had already had some experience in building trestle piers in the Shetland Isles to enable stores to be unloaded for RAF bases and he appreciated how unsuitable they could be in rough seas.

After a search along the British coastline, a suitable development area was found in the Solway Firth where the rise and fall of the tide was similar to that found off the Normandy coast and the area round the small fishing village of Garlieston was sealed off by barbed wire from the rest of the world. Carline rapidly established friendly relations with the local fishermen and turned a blind eye to their officially-prohibited activities with the proviso that the occasional gift of salmon or lobsters would not be unwelcome for the mess. In the event, the security of the locals proved to be first rate, though Carline did not appreciate this until after the war was over. When visiting the War Office, a colleague informed him that an officer wished to return his hospitality and he was approached by a colonel with whom he was unfamiliar. 'You once treated me to a drink,' declared the latter. He had, it transpired, been sent by MI5 to penetrate the prohibited area at Garlieston in order to discover the state of security. Leaving the station, he had dressed as a tramp and walked the forty or more miles to the coast. He was sitting in the harbour bar when Carline walked in accompanied by two American officers. After standing them drinks Carline noticed the dishevelled figure in the corner and ordered the publican to 'give him a pint'. The MI5 man failed to discover what was afoot at Garlieston. Such was the high level of security during the construction of the Mulberry components and, considering the thousands of workers who, at one time or another were involved, it was remarkable, that the enemy had no inkling of what was afoot.

In such a remote part of Scotland recreational facilities for the men of the development unit were limited. The all-ranks football team soon found that their ability did not match that of the local teams. Steer-Webster, on learning this, at once located a number of Scottish First Division players scattered over various theatres of war. They were flown home at his request and sent to Garlieston to raise the sappers' morale and in a short time the opposing teams were routed, to the great delight of the army.

By the autumn of 1942 specifications for a pier and pierhead were put forward by Mountbatten. The pierhead had to be capable of berthing three 2,000-ton coasters, the pier had to allow a continuous flow of traffic and it would have to be not less than a mile long. Three schemes were submitted and it was the responsibility of Tn5 to discover which was the most suitable.[2]

The first was proposed by Iorys Hughes, a consulting engineer who had been responsible for the design of the Empire swimming pool at Wembley and who was also a well-known yachtsman. His design consisted of a fixed pier and pierhead. There were two components, a large concrete caisson called a Hippo which carried a tubular steel span called a Croc. The Hippo would be capable of being towed across the Channel and sunk in position, the Croc subsequently being laid on top to form a roadway.[3] Bruce White decided that a prototype should be built by Holloway Brothers, a firm of construction contractors. They had already built, under the direction of their chief engineer, W. Storey Wilson, several forts, designed to carry anti-aircraft guns and radar equipment, attached to concrete pontoons, which were then floated into the middle of the Thames estuary and sunk.

No dry dock being available, as for the forts, a building site was chosen on a secluded golf course at Conway, North Wales, which sloped down to the tidal waters of the estuary. Storey Wilson decided that the caissons should be launched sideways from a slipway, one behind the other. Such large craft (each caisson weighed some 3,200 tons) had never been launched in that way before.

The second proposal came from DMWD and was the invention of an engineer named Ronald Hamilton.[4] Prevented from entering the navy on account of a crippled arm, Hamilton had, since early in the war, become preoccupied with designing floating structures based on the principle of flexibility as opposed to rigidity. He had already designed a floating airstrip built out of a mosaic of hinged hexagonal tanks filled with air.

Swiss Roll, as his proposal later became known, was a floating roadway built of wooden planks held together with steel cables. The principle was that a load passing over it would be held partly by tension in the cables and partly by normal Archimedean displacement. A vehicle driven along the road raised hinged canvas or wooden sides, thus preventing water from flooding it. Swiss Roll was intended to be attached to a floating jetty built of hexagonal units like the landing strip and had the great advantage of being light and cheap to make. But it was unable to carry loads greater than seven tons, so that bringing tanks ashore was out of the question.

The third design was sponsored by Tn5 itself. This was a floating bridge cum pierhead, the latter being fitted with legs or spuds to enable adjustment to be made according to the state of the tide.[5]

William T. Everall, who designed the spans of the bridge, was a railway engineer who had acquired an almost legendary reputation in north-west India. He had returned home shortly before the war to become War Office adviser on railway bridging and in September 1939 he was appointed Chief Bridging Instructor, No 2 Railway Training Centre, Derby. After the fall of France, the War Office anticipated that on returning to the continent both port facilities and rolling stock would have been destroyed by the enemy. Everall was therefore asked to design a flexible gangway to enable locomotives to be loaded and unloaded from a Channel ferry steamer. Using spherical bearings, Everall made it possible for a span to twist through an angle of 6 degrees to conform to the motion of the vessel. (See Fig 3.)

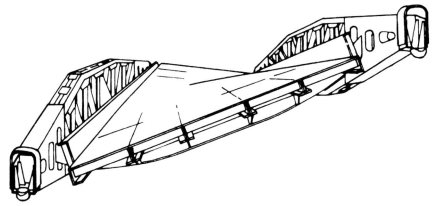

Fig 3 Flexibility of standard span

The bridge spans were 80ft long and were made of mild steel, welded and black-bolted in construction. (Black bolts are bolts covered with black iron oxide called scale, they are of less uniform shape than bright *turned* bolts.) Each span weighed 28 tons. Everall used spherical bearings to enable the span carried by a float to twist when riding on a rough sea. They allowed a free angular movement of one span relative to another of 24 degrees together with a torsional displacement of 40 degrees along the length of each span. (See Fig 4.) Allowance

Fig 4 Arrangements of bearing for floating roadway

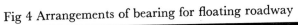

for these large angular movements was made by arranging for the spans to sit one inside the other, the outer bearings only taking their support on the float. A third connection, a link from the centre of the end girder, arrested the rolling tendency of the float.

A telescopic span was also designed. Like the standard span it was flexible, but had a sliding panel which allowed the span to be lengthened or shortened between 71 and 80ft. The span, which was introduced at intervals along the roadway, ingeniously overcame four problems. Firstly, it enabled the bridge to adjust itself to variation in the height of tides, especially spring tides some of which were higher than others, without having to be pulled up the beach and thus making it necessary to close the bridge for a period. Secondly, the mile-long length of the bridge caused a variation in the angle of the roadway at high and low tides. (See Fig 5.) When in the low-tide position the bridge would be slightly longer than at high tide. The telescopic spans allowed the bridge to expand and contract according to the rise and fall of the tide. Thirdly, a choppy sea would not only cause the bridge sections to rise and fall but also to move sideways to and fro. When the bridge moved out of a straight line it naturally became slightly longer. The telescopic spans allowed for this expansion and contraction and saved the flexible couplings joining the spans from tremendous strains and stresses. Fourthly, the telescopic spans were used to link one pierhead to another. This avoided exact placing of the latter as the telescopic spans could adjust themselves to the precise distance at which the pierheads happened to be.

The main girders were lozenge-shaped so that the chord material was well-disposed to withstand bending stresses and longitudinal forces. Tensile forces along the bridge were resisted by steel cable strops which were secured to the bearing housing after passing through a very large sheave worked into the girder end.

The roadway, 10ft wide, was composed of steel decking supported on cross girders and a special non-slip tread was devised for it by Everall. The centre cross girder was 3ft deep and had a bolted connection at each end so that it held the planes of the main girder parallel. The top flange was of

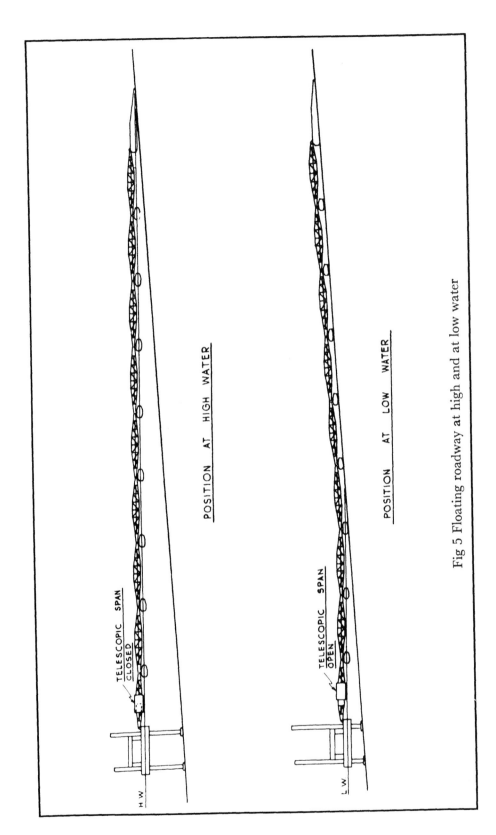

Fig 5 Floating roadway at high and at low water

sufficient width to provide a seating for the deck units. There was no bottom flange but steel plates were riveted along the bottom edge to give a certain amount of bottom chord area. It was able to accommodate in its length about 5 degrees of twist without overstrain. The other cross-girders were pinned to the main girder by a $2\frac{1}{2}$in-diameter pin so that they offered no restraint to the twisting of the bridge.

Under Everall's supervision, Messrs Braithwaite of West Bromwich developed the standard and telescopic spans and helped to design the steel roadway, while John Thompson of Dudley's developed the components forming the end of the spans. The box construction and bearing attachments were all welded. Welding (a method hitherto largely shunned by British structural engineers) as opposed to riveting, was carried out whenever possible as it saved many tons of steel, then in such great demand. Tn5, having no experimental establishment, most of the design work took place at the Carriage & Wagon Department of the London Midland & Scottish Railway at Derby. A full-scale wooden model of the bearings was made and this helped to solve the problem of the amount of clearance required under various positions of the bridge.

During the early part of 1943, Everall was attached to the British Army Staff in Washington, so that much of the early development work fell on the shoulders of his principal assistant, Allan Beckett.[6] Beckett had worked with Braithwaites and, as will be seen, he also made original contributions of his own to the bridge equipment. Not content with development, he insisted on accompanying Whale to Normandy. He was then holding the rank of major in the Royal Engineers. As the development progressed, most of the co-ordination took place in several offices in County Hall, London, on the South Bank in close proximity to Tn5 in Whitehall across the river.

The floats carrying the bridging required a great deal of attention. Each one had to support a total weight of 56 tons, including a 25-ton tank. They had to withstand being towed sideways for a long cross-Channel voyage and they had to harmonise with the spans they carried when anchored in a rough sea. (See Fig 6.) Finally, they had to be mass-produced cheaply and easily. Everall designed a steel float which fulfilled

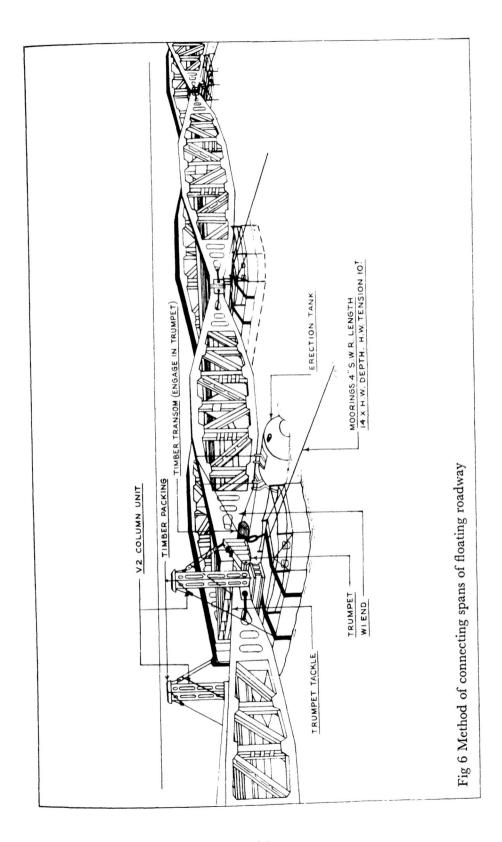

Fig 6 Method of connecting spans of floating roadway

most of these requirements. It was ellipsoidal in shape, 42ft long, 15ft wide, 8ft deep and was formed of six separate all-welded units bolted together. The stiffened steel plating was $\frac{3}{16}$in thick but $\frac{1}{4}$in thick at the bottom where it would have to rest on the sea bed when the tide was out. The total weight was 16 tons. The steel plate was designed to withstand $\frac{1}{4}$ ton per square foot. Most of the development was done at Thompson's.

It was calculated that about 460 floats would be needed for the Mulberries and that if built in steel they would absorb some 5,000 tons. But as 60,000 tons were already earmarked for pier-heads and bridging, to construct all the floats in steel would have seriously jeopardised the shipbuilding programme, particularly escort vessels to counter the U-boats.

The alternative was to make most of the floats in concrete. Following the example set in World War I, concrete-built vessels, mainly petrol barges, were already in use. Cyril Wood of L. G. Mouchel & Partners, the well-known consulting engineers, had even designed a concrete cargo vessel, *Armistice*, of 2,500 tons which, in the inter-war years, had plied between the United Kingdom and the west African coast. Tn5 therefore asked Mouchel's to design a series of experimental concrete floats.[7] They were to be built by Messrs Wates, one of the few building firms then working in pre-cast concrete, at the Vickers shipyard at Barrow-in-Furness where there were suitable handling and launching facilities.

Three problems had to be solved. The weight of the concrete had to be reduced to make the float easy to tow; the skin had to be waterproof; and some form of fendering was necessary to prevent it from being holed.

The first concrete floats were too heavy but, after a number of experiments at Barrow, Wates succeeded in producing panels no thicker than $1\frac{1}{4}$in with stiffening ribs. They formed a float, boat-like in shape, which was divided into six watertight compartments with transverse bulkheads 2in thick. The latter took shear stresses. Each compartment had a manhole through which it could be inspected. Wooden fenders protected the sides and the bottom from being damaged, particularly the latter when the float rested on the sea bed. As it was decided to tow the floats sideways, they were given a turtleback similar to the steel floats in order to reduce resistance. For this reason they were called Beetles (officially pier pontoons). The turtleback also reduced the pitching tendency to which the prototype floats were prone. (See Fig 7.)

The Beetles did not evolve overnight. A combination of modified Thames barges and concrete floats shaped like barges with a displacement of eighty-seven tons were used in the initial trials and a succession of experiments was necessary to find out the best design for towing and for carrying the spans over long distances at sea.

The pierhead for the Tn5 design was based on a dredger called *Lucayan*, designed in 1923 to work in the Bahamas and built by the Renfrew shipbuilding firm, Lobnitz & Co.[8] The vessel had to be kept stable, so she was fitted with three legs which were embedded in the sea bottom and on which she was able to slide up and down with the tides, not moving meanwhile. During a storm she survived while other craft came to grief. A spud pierhead would have a great advantage over a fixed or anchored pierhead in that coasters could discharge at all states of the tide.

Bruce White consulted Henry Pearson Lobnitz, managing director of the firm, about the practicability of the scheme. Lobnitz found no difficulty in adapting the dredger and spuds

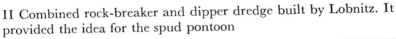

II Combined rock-breaker and dipper dredge built by Lobnitz. It provided the idea for the spud pontoon

38

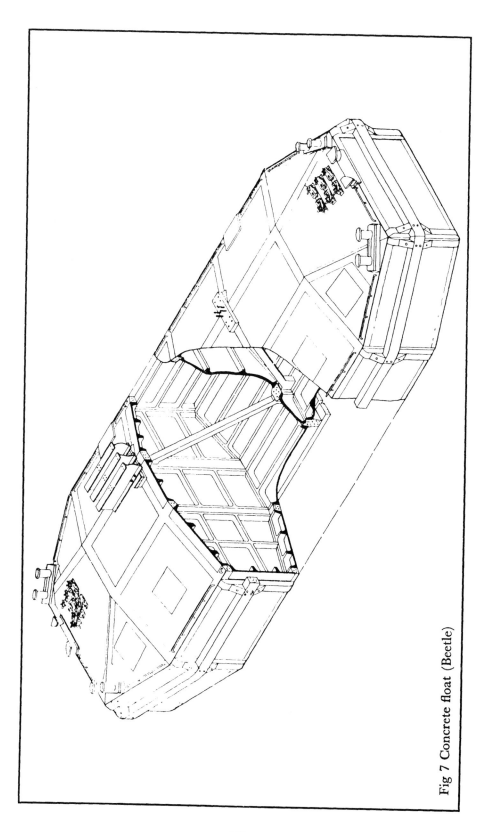

Fig 7 Concrete float (Beetle)

39

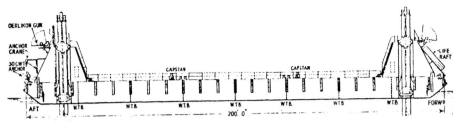

Fig 8 Arrangement of spud pontoon

to a large steel pontoon and in a matter of weeks a design was ready. It had a very distinctive shape.[9] At each end there were two high stacks like chimneys. (See Fig 8.) These housed the spud legs which were no less than 89ft long with feet 4ft square. They were contained within guide frames enabling the spuds to slide up and down but giving great lateral stiffness. The spuds were operated by twin wire cables attached at one end to a 20hp electrically-operated winch with a reduction gear, and, at the other, to the top of the spud. The downhaul cable which passed over a sheave at the top of the spud lifted the hull by pressing the spud foot down on to the sea bed. The other cable passed over the lower end and was used for lifting the spud. (See Fig 9.) The spuds could be raised or lowered at a rate of approximately 2½ft a minute. The winch motors were fitted with a magnetic brake which was set to hold a 12in lift of hull corresponding to a load of 84 tons on the spud. If this load was exceeded the brake automatically released the load to this figure. The motive power for the winches was provided by two 57kw diesels and generators, which also operated bollards, bilge pumps and other equipment. One diesel was usually sufficient to operate the spuds; the second came into action when the weather demanded extra power.

When floating with the spuds raised the pontoon drew about 3ft 3in of water. When the spuds were lowered the draught was approximately 2ft 10in; the hull was then relieved of the weight of the spuds, each of which weighed some 35 tons. In calm weather the pontoon floated up and down with the tide, the spuds providing sufficient anchorage. In the normal operating position the hull was raised 6in above the free floatation level. This imposed a load of 42 tons on each spud. In heavy weather the spuds were supposed to lift the pontoon at least 12in. But

40

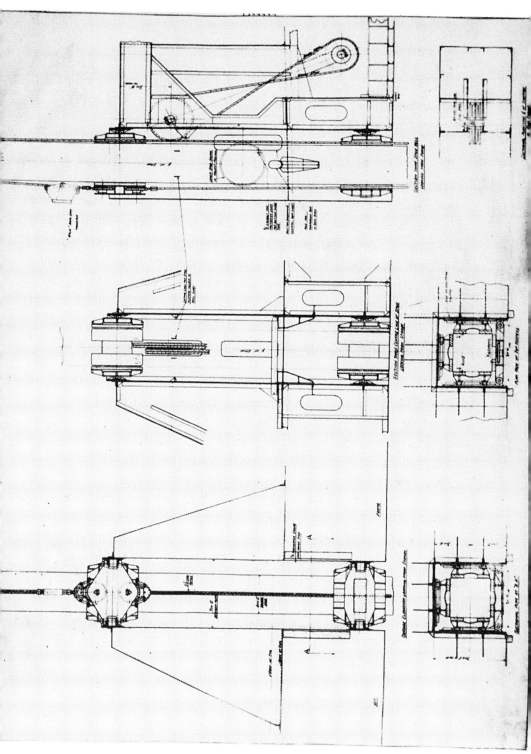

Fig 9 Operation of spud legs

they were not designed to lift or hold the pontoon completely out of the water. (Plate VII.)

The pontoon itself, like a small ship, was 200ft long, 60ft wide and had a moulded depth of 10ft at the sides. The total weight, including spuds, was around 1,100 tons. Internally, it was divided into watertight compartments and there was room to accommodate the operating crew of one officer, six NCOs and fifteen men. It was equipped with kitchen, sanitation and heating appliances.

As Lobnitz was unable to undertake construction of the prototype, this was entrusted to Alexander Findlay & Co of Motherwell. Primarily structural engineers and bridge builders, they had been switched to the building of tank landing craft, flak ships and gunboats. They had converted for this purpose a derelict shipyard on the Clyde and here the pierhead was to be constructed.

The target date for the trials of the three designs which have now been described was to be March 1943. Meanwhile, the conception of the floating pier was never absent for long from the Prime Minister's mind and, in fact, the development of Mulberry continued to hold his attention until the end of the war. He even kept a working model of the scheme, prepared by Tn5, next to the War Room in Whitehall. Early in March he was already fretting that the 'matter is being much neglected. Six months [have elapsed] since I urged construction of several miles of pier.'[10]

Four days earlier, in fact, six spans of the Tn5 project had been loaded into three LCTs at Barrow-in-Furness for shipment to Garlieston. The floats, towed by tugs, had already arrived, closely followed by a 60-ton floating crane. The task of linking the spans at Cairnhead, near Garlieston, was given to 167 Railway Bridging Company, RE, under Capt (later Lt-Col) D. J. Tonks, who became responsible for the follow-up of the Whale components after D-Day and later assumed command of 970 Port and Floating Equipment Company, RE, at Arromanches. Work began on 17 April and was completed on the 22nd.

During this stage the method of erecting the spans was complicated. At the harbour the crane hoisted two spans, one above the other, onto two floats placed abreast. (The floats also were equipped with lifting gantries.) Known as 'Grasshoppers', the floats were towed abreast to the trial site. On the shore a trestle ramp awaited connection with the spans. When the Grasshopper was in position the top span was slid forward by means of a rubber-mounted trolley and hoisted into position by the gantry. The two spans were then linked together. The process was repeated with the next two spans and so on.

The erection team found it very difficult to control the movement of the lower span relative to the top one in spite of using arrester gear. After a number of experiments an entirely new method was devised involving the use of an erection tank

III Linking spans of the floating roadway, using erection tank (*Public Record Office*)

to link one span to another.[11] The tank was of welded steel, cylindrical in shape, 8ft in diameter and 6ft long. The erection tank, filled with compressed air, carried the end of the span while being towed. When in position the tank was flooded, allowing the ends of the adjacent spans to be aligned. Trumpet-shaped guides then enabled the spans to be linked together rapidly. The tank was then towed away. The erection tank could also be used when damaged floats had to be replaced. (Plate III.)

The stability of the bridge depended on the system of mooring the floats. They had to be kept within a range of movement of about 6in irrespective of wave action and the rise and fall of the tide. It would be impossible to maintain a flow of traffic when the floats were moving independently of the roadway.

For the initial trials, mooring wires were attached from each pontoon to concrete clump anchors, each weighing 5 tons and kept taut by winches on the floats. They had to be laid by special shallow-draught ships equipped with derricks, a procedure which would clearly delay erection of the roadway. An anchor was therefore required which was strong, yet not too

IV Prototype pier and pierhead at Cairnhead

V Allan Beckett's kite anchor used to moor floating roadway

heavy, and which could bury itself in the sea bed as the load on it increased. Beckett experimented with, and finally developed for use where there was a sandy bottom, a kite anchor which in some respects resembled the naval CQR anchor used for mooring LCTs. It was made of steel and fitted with a shoe, or plough. When the anchor dragged, the point of the plough dug into the sea bed making a trench. A ship equipped with this anchor could immediately be brought to a standstill.

Special equipment was devised for dropping the anchors and laying the mooring cable. Firstly, there was a raft formed by two long floats carrying a drum on which was wound 1,200ft of mooring wire. The wire could be prevented from running out too fast by a hand brake. Two kite anchors were tied in position at each end of the raft. The raft, or shuttle as it was called, was towed by a shallow draught, twin-engined motor boat designed by Camper & Nicholson Ltd, the well-known yacht builders. It was called a Surf Landing Under Girder (SLUG) Boat for a reason which will soon become apparent. The shuttle was carried on one of the spans and launched from a temporary ramp, using the drum as a wheel. It was then

VI Mooring shuttle which carried the cable for securing the floating roadway

towed by the SLUG boat to the upstream anchor position about 600ft away from the pier. The cable was attached to the anchor, which was then cast. The shuttle was then towed to the downstream anchor position paying out wire as it went. The SLUG boat was low enough in the water to pass *under* the bridge and after connecting up again, dropped the second anchor in position. The cable meanwhile had been lifted out of the water at the bridge by a bight of wire and made fast to the float. The cables were made taut by a Yale 'pull lift' and a rope stopper which were kept on the floats. The 'pull lift' was first attached to the float transom and then to the stopper, after the latter had been clamped to the mooring wire. The wire was then hauled in by the 'pull lift'. This method eliminated the necessity of fitting all the floats with winches—an item of great expense. In rough weather it was particularly important to ensure that the wires were not slack. The SLUG boats, which had a crew of two, were 20ft long and weighed 2 tons, could also carry a winch for kedging or compressors for de-watering the erection tanks. They were carried on the shore ramp float, which was designed to connect the floating bridge to the shore. It was 80ft long and made of thin steel plate in four separate all-welded units black-bolted together. The float was submersible and the seaward end carried the end of the first bridge span.

The early trials also showed that the floats nearest the shore would be vulnerable to damage from protruding rocks. For this reason only steel floats were to be used. These were fitted with

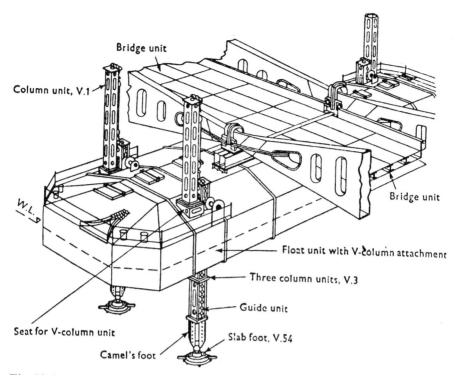

Bridge unit

Column unit, V.1

W.L.

Bridge unit

Float unit with V-column attachment

Three column units, V.3

Guide unit

Seat for V-column unit

Slab foot, V.54

Camel's foot

Fig 10 Steel float with adjustable legs for grounding on rocky foreshore

miniature spuds and operated by a simple hand wheel locking device which enabled them to be adjusted at various depths. When the floats were moored, the legs were let out and locked, leaving the float and spuds clear of the beach when the tide receded. Compression tools then sheared off the jagged edges of rock and a bed of sandbags was laid. When the tide came in, the legs were raised to their uppermost position, after which the floats grounded on the sandbags. (See Fig 10.)

Meanwhile, on 23 April, the pierhead, completed in the remarkably short time of four months and christened *Winnie* by her builders, arrived from the Clyde in Wigtown Bay. A strong south-east wind was blowing, but the spuds were lowered and after four days of bad weather, the pierhead was connected to the bridge spans.

In a few weeks the equipment underwent its first spell of rough weather.[12] Strong winds gusted up to 60mph from time to time and the waves rose to heights of six and occasionally twelve feet. Small craft in the bay were driven ashore and smashed, but the floating structure held despite the dragging

VII Pierhead being carried by its spuds (*Public Record Office*)

VIII Intermediate pontoon linked to spud pontoon to form enlarged pierhead

of the mooring clumps. Apart from a few loose bolts in the decking of the roadway there was no sign of failure.

On 18 June there was a demonstration before the Chiefs of Staff and the Engineer-in-Chief, Maj-Gen C. J. S. (later Sir Charles) King.[13] A 10-ton tank, a tank transporter and 10-ton lorries were driven along the bridge. Although the sea was calm, the spectators were impressed, including even, as Beckett noted wryly, the Admiralty representatives. He also observed with satisfaction that the long taut cables kept the floats steady. Without them, trucks would not have been able to drive at 40mph as if down a main road.

Spud pontoons, operating singly, could not berth more than one ship at a time. A proposal was therefore made for several spud pontoons, about 160ft apart, to be linked by intermediate pontoons, 80ft long by 60ft wide, attached to one spud pontoon and a telescopic span linking the free end of the intermediate pontoon and the next spud pontoon. The intermediate pontoons had flared sides and swim ends to make them easy to tow. Each one was divided into eighteen compartments, of which three were reserved for stores and one for living accommodation, if required. They were linked by cables and strops to spud pontoon and telescopic spans. As they provided a space for the unloading of stores, the decks had to be strong enough to carry heavy lorries but they were not intended to resist berthing shocks. The spud pontoons were protected by birchwood fenders and kept vessels from berthing against the intermediate pontoons. (See Fig 11.)

The shortage of steel made it necessary for these units to be built of concrete and, like the floats, they were designed by Mouchel's. However, a number were, in the end, made of steel as the concrete sides were so vulnerable. To suit existing templates and jigs, the intermediate steel pontoons were similar in construction to the middle sections of the spud pontoon and the whole arrangement became known as the Stores Pier.

Further trials, however, showed that lorries could be driven off an LST on to a ramp built on the pierhead in about five minutes. Admiralty officers watching the demonstration asked that a separate pierhead be developed for discharging LSTs. These craft had a forward draught of 4ft 6in and 11ft 6in aft

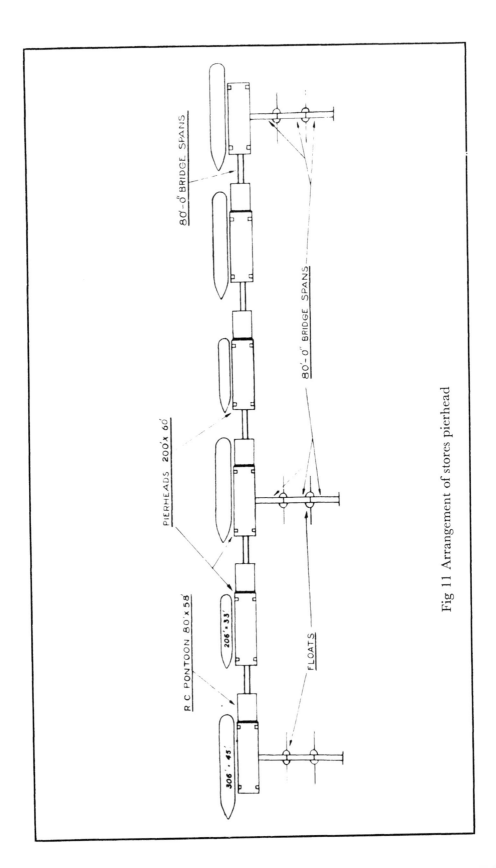

Fig 11 Arrangement of stores pierhead

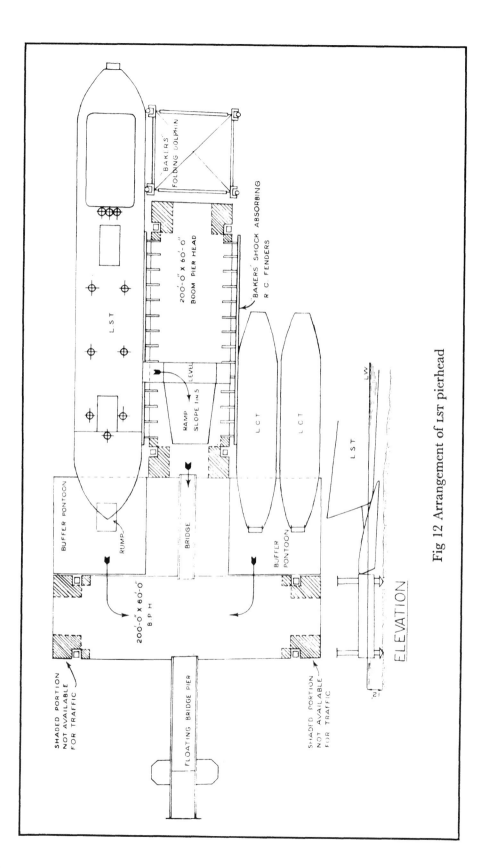

Fig 12 Arrangement of LST pierhead

when fully loaded and when vehicles emerged from the bow doors they had to be driven through water to reach dry land. The vehicles on the upper deck had to be lowered by lift after the lower deck had been cleared. Unloading could take up to an hour and a half.

Tn5 now designed a special LST pierhead and the officer responsible was Richard Pavry, who had formerly worked on bridging problems with Dorman Long.[14] Two spud pontoons would be placed in the form of the letter T and connected to each other by a telescopic span. One pierhead would be positioned in line with the bridge while the other would be set at right angles to it. If the LST could be run up an artificial ramp fixed to the head of the T, vehicles could unload from both decks simultaneously, those from the lower deck coming straight out onto the head pontoon, while those on the upper deck could be driven down a ramp onto the second pontoon. (See Fig 12.)

As may be seen, the T shape enabled two LSTs, or LCTs, to be discharged at the same time. This simple scheme enabled a very much more rapid discharge of vehicles than had been hitherto possible.

But heavily loaded LSTs of 2,557 tons dead weight approaching the pierhead at a speed of 3 knots required a buffer of very special design. A pontoon was therefore designed by Storey Wilson assisted by F. W. Sully, also of Holloway's and who did

IX Launch of buffer pontoon at Conway

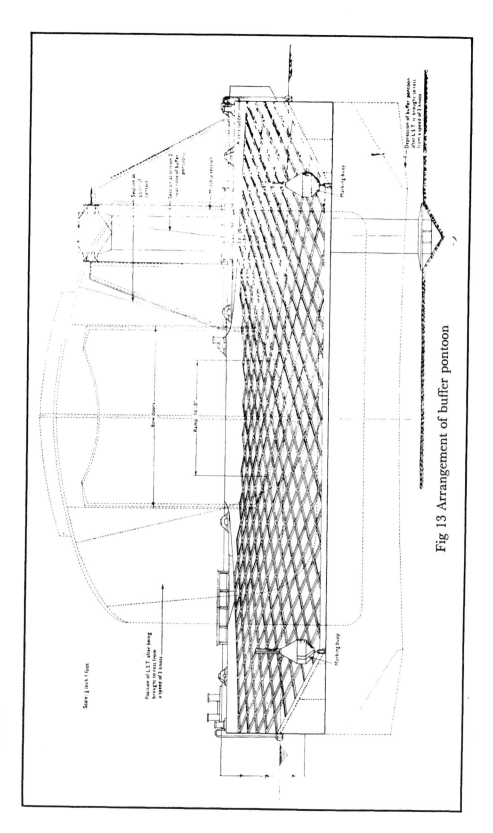

Fig 13 Arrangement of buffer pontoon

the major part of the calculations. The pontoon was shaped like a ramp, the thick end being hinged to the spud pontoon. The hinges, which had to be very strong, were made of steel rollers and were attached to hardwood bearings.

When berthing, the bows of the LST pressed down on the sloping deck. A number of experiments were conducted at the National Physical Laboratory to discover the correct incline and shape of the deck. It had to be dish-shaped so as to give the maximum water plane. Wilson and Sully decided to design the deck in the form of an equilateral hyperbolic paraboloid. This provided the most suitable curve for water plane areas and produced a troughing effect, enabling the LST to be guided towards it. The design also had the advantage of simplifying manufacture for all the lines on the surface of the paraboloid at 45 degrees to the axis were straight. The interior of the pontoon was divided into cells. The top edges could therefore be made straight, each cell being boundaries to a plane surface. The sides of the cells could therefore be set out conveniently in straight lines and flame-cut on site. (See Fig 13.)

Another problem was the need to cushion the effect of LSTs berthing alongside the pierhead.[15] Special fenders were proposed by A. L. L. Baker (later Professor of Civil Engineering at the Imperial College of Science and Technology). Earlier in the war Baker, an authority on concrete engineering, had designed a suspended fender for oil tankers berthing at a recently-constructed, but exposed, jetty at Heysham on the Lancashire coast. But this cumbersome attachment could not only damage the structure of the pontoon but could also give trouble during a voyage. Pavry therefore concluded that it would be better to make smaller units, about 2–3ft long, and which were capable of being secured. He asked his friend Ronald Jenkins, a brilliant structural engineer with considerable mathematical ability, to work out details for a method of transferring the load from one unit to the next. Jenkins and Ove Arup (now the celebrated structural designer) with whom Jenkins had collaborated for some years designed a fender 2ft long and weighing 2 tons, crank-shaped so that when it was pushed back and upwards the ship's side would not bear on the bracket. Screwed rods were passed through sleeves in the brackets and hooked

X A. L. L. Baker's gravity fenders attached to spud pontoon

into the back of each unit, so preventing the sides of the pontoon from being damaged during towage. (Plate X.) The maximum lift of the units when pushed back to their extreme position was 20in. They extended 150ft along the sides of the pontoon and increased its weight by 300 tons. (See Fig 14.)

Thus the conception of the LST pierhead was simple, but the individual parts required much thought and experiment. An unexpected side-effect of the Baker fenders was the unending screaming noise as they ground together along the side of the pontoon.

Baker also designed large, collapsible floating dolphins made of long steel tubes, which were to form an extension to the pierhead and to secure the LST's stern.[16] They could be folded rather like a deck chair for towing to the site and then vertically erected with the legs on the sea bed.

Trials of the two other pier and pierhead schemes followed in due course. The Hughes fixed pierhead and spans was installed at Rigg Bay, near Garlieston. But the main disadvantage was that it could not rise and fall with the tide (though Storey Wilson has pointed out that adjustable approach spans could have been made and were, in fact, designed for the Nanking–Pukow train ferry in China). Apart from this, scour caused the

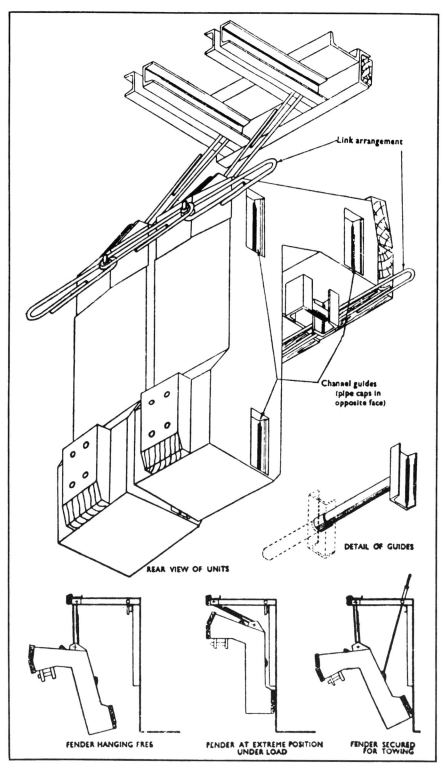

Fig 14 Arrangement of gravity fenders

pier to fail. But the work on the Hughes pier was far from wasted as much of the technique in building the concrete caissons was applied to the breakwater units.

Swiss Roll, on the other hand, although cheap and easier to build than the other two proposals, had the great disadvantage of being unable to carry tanks or heavy vehicles. A number of trials were made that summer at Cairnhead with the equipment, including running various types of motor vehicle over it. Usually the engine of the vehicle, which could only be driven slowly in second gear, became overheated. Swiss Roll also did not survive the rough weather. However, 2,000ft of Swiss Roll were laid at Arromanches in the Mulberry and used briefly for the disembarkation of naval personnel.

By early July, 1943 the requirements for Overlord had been enumerated and the Chiefs of Staff had no alternative but to approve the Tn5 scheme.[17] They ordered four miles of pier and six pierheads, which were to be ready by 1 February 1944.

3 The Need for Sheltered Water

The pier and pierhead were originally intended for discharging supplies off an open beach without the benefit of sheltering breakwaters. Proposals for Overlord, however, envisaged nine divisions in the assault to be followed by twenty divisions for building up the assault area. During the first month of operations on the continent eighteen divisions, and during the second month, twelve divisions would have to be maintained over the beaches. This would include the unloading of supplies, ammunition and motor vehicles. If their discharge was to proceed without interruption in the predictably choppy Channel sea, some form of sheltered water would have to be provided.

Hence the idea of an artificial harbour was born. Exactly who proposed the idea is disputed. But the most likely person was Commodore (later Vice-Admiral) John Hughes-Hallett, RN, naval commander for the Dieppe raid, and afterwards Naval Chief of Staff to the Overlord planners. Legend has it that at a meeting he declared that as a port could not be captured, it would be necessary to take one across the Channel; a remark greeted with laughter. Certainly by June 1943 the proposition was being taken seriously when Hughes-Hallett and Maj-Gen C. N. (later Sir Nevil) Brownjohn, in charge of administration on the Overlord planning staff, attended Conference Rattle.[1] This important inter-service meeting held in Scotland was convened by Mountbatten to discuss outstanding technical and administrative problems which would arise in Overlord. Hughes-Hallett suggested making a breakwater by sailing to the site seventy obsolete merchantmen and sinking them—a standard naval solution. Brownjohn suported this idea, but argued that equal attention should be given to floating equipment for the discharge of supplies.

Proposals for breakwaters, stimulated in the first place by the loss of Singapore, had come from COHQ, principally from Major C. La R. Salter, an officer in the Royal Marines.[2] The first proposal was for a form of wave suppression using compressed air. An American engineer, Philip Brasher, had suggested this idea and it had been tested by the Standard Oil Co at El Segundo, California in 1916. A perforated pipe laid just above the sea bed ran in an arc from the shore enclosing the area to be treated. Compressed air was pumped through the pipe and emerged through the holes, thus creating a sufficiently calm area of water to enable vessels to discharge at a pierhead in poor weather. A similar system had been used to salvage the USS *Yankee*. In 1925 the Admiralty had displayed some interest in the invention but had done nothing about it.

Goodeve had initiated experiments at the City and Guilds Laboratory in South Kensington. They were conducted by Dr C. M. White, attached to the Technical Section of DMWD. Little interest was taken by senior officers in COHQ on the grounds that development would take too long to be of any practical use in the war.[3] Nevertheless at the end of February 1943 Goodeve, who was now Assistant Controller Research and Development in the Admiralty, urged that the development of artificial breakwaters should proceed on the same priority as that given to the pier and pierhead.

The compressed air scheme was now taken over by an electrical engineer named Robert Lochner, in charge of the Heavy Engineering Section of DMWD.[4] Like Hughes and other officers involved in Mulberry, Lochner was an enthusiastic yachtsman. At the beginning of the war he had been involved in the development of counter-measures against magnetic mines known as 'degaussing'. As he was unfit for seagoing duties on account of poor eyesight (although he had been in charge of a minelayer for a while) Goodeve recommended him to DMWD.

After model experiments, full-scale trials took place. At the DMWD experimental centre at Weston-super-Mare, known as HMS *Birnbeck*, compressors and perforated pipes were put into operation. But Lochner and his colleagues quickly discovered, if it had not been obvious from the start, that an enormous quantity of generating power would be required, while a heavy swell would soon put the pipes out of action.

From *Birnbeck*, therefore, the experiments moved to Portslade, the industrial end of Brighton, in Sussex, where, conveniently situated near the beach, there was a power station. One windy day in the spring of 1943 Lochner called on the Brighton Borough Electrical Engineer to ask him for a beach, a supply of electricity and for the use of his workshops. 'At the same time', wrote Lochner afterwards, 'I was quite unable to explain why we wanted these things. I need have anticipated no difficulty. He not only gave us all we wanted, but was the soul of discretion, asking no questions and not showing by a flicker of an eyelid any hint of our purpose. But I think he made a pretty shrewd guess.'

A bubble breakwater, some 1,200ft long, was laid 600ft out to sea. But again the compression and equipment required presented immense problems. A prolonged spell of bad weather made matters even worse. 'Pipes went adrift', wrote Lochner, 'or were silted up, weights and anchoring tackle would be picked up and deposited in a heart-breaking tangle on the beach.' Reports of shortages of equipment came in. But even 'assuming that we were able to obtain the machinery in reasonable time and even taking the lowest figure of total horse-power which we could safely allow, the problem of embodying these pipes and this vast quantity of machinery in a portable layout was a nightmare.' Moreover, as the sheltered water at Portslade was only half the area of what would be required, it was calculated that at least $1\frac{1}{2}$ million horse-power would be required to calm the water day and night.

Even so, the Admiralty persisted with the bubble breakwater long after its impracticability had been demonstrated. Further experiments took place at Cairnhead, but again were a failure because of the lack of compressing power. The scheme was finally abandoned at the end of 1943, lamented, perhaps, only by Churchill who had a predeliction for improbable schemes of this nature and who continued to inquire, petulantly, about the progress of 'my Bubbles'.

Lochner, who had an inventive mind, had, meanwhile, thought more deeply about the problem. Recovering from influenza, he recalled experiments in connection with the bubble breakwater at the Admiralty experimental tank at Haslar, near Portsmouth. Wave molecules were shown to move

XI Robert Lochner experimenting with the first Lilo model on his trout pond

in a circular path. At the same time it was obvious that waves were only surface deep. Light began to dawn. Before long Lochner had evolved the following two principles. Firstly, that to interrupt a wave the interrupting barrier need only go to a depth equal to about 15 per cent of the wavelength. Secondly, that a wave can be interrupted and reflected by means of a floating barrier, provided that each of the natural periods of the floating barrier was longer than the period of the longest wave to be stopped.

Forgetting that he was convalescing, Lochner put these principles into practice. He bent a rubber Lilo, last used on a summer holiday before the war, lengthways. Enclosing some gas pipes to form a keel he asked his wife to sew the two sides together. The two of them then launched this strange-looking object on the trout pond at the bottom of their garden. Lochner observed the reaction of the Lilo to waves made by his wife with a biscuit-tin lid. He believed he had stumbled on the solution to a breakwater which would be easy to build, cheap and mobile Such a flexible floating barrier would eliminate the machinery, pipes and ships that would be necessary to put the bubble breakwater into operation.

Filled with enthusiasm Lochner changed his clothes, rushed up to the Admiralty and worked out designs for some more accurate models. They were tried out at the City and Guilds Institute. 'On the model scale there was no doubt left. The principle of suppressing waves by means of a floating barrier appeared to be fundamentally sound.'

Lochner was now instructed to organise the project known first as 'Lie-low', later shortened to 'Lilo'. The work was planned so that its various aspects could proceed simultaneously and, to some extent, independently. Dr W. G. (now Lord) Penney, then Assistant Director of Mathematics at Imperial College, and Mr A. T. Price assisted with mathematical theory. Lieut Urwin, RNVR, lately of the Public Record Office, was in charge of models at Haslar and Dr E. A. Guggenheim of Reading University analysed these experiments. Mr H. Bateman and Mr Dyer of the Balloon Establishment at Cardington designed the vast air bags. Bateman had devoted much of his life to airships and was one of the survivors of the R38 which broke in half over the Humber on a trial flight. Major S. Nixon, Deputy Director of Balloon Production in the Ministry of Aircraft Production, found the materials for manuacture. Dr Oscar Faber, the reinforced concrete expert, who had worked in the Admiralty in World War I, designed the concrete keel to weigh down the balloon.

By July the mathematical theory had been completed, over 100 model experiments performed, full-scale designs roughed out, and the floating dock at Portsmouth cleared for construction of three full-scale prototypes. The unit looked like an early rigid airship without controls. It was 200ft long, 12ft wide and had a draught of 16½ft. It consisted of four three-ply canvas envelopes (the first of cotton was unsatisfactory) placed one inside the other and enclosing bags to be filled with air, each running the full length of the unit. The envelopes were attached to a 700-ton reinforced concrete keel. The air pressure was adjusted to coincide with the mean hydrostatic pressure on the outside of the respective envelopes. In that way a form of hull side was obtained which moved in or out under any temporary imbalance between these two pressures corresponding to any alteration of the immersion depth. Consequently the restoring force with that type of hull was only a small fraction of that for a rigid-sided hull if the same displacement and the periods of the wave motions were correspondingly lengthened. Construction of two prototypes began in July.

Obviously this long, inflated envelope floating on the sea would be extremely vulnerable, not only to sharp floating objects which would be liable to puncture it, but also to deflation by enemy action either from the sea or from the air. Lochner therefore pursued his researches with the object of achieving a rigid or hard version of Lilo. It was now mid-July and Lochner decided to put into practice one of his constructional principles, namely the enclosure of a large mass of water within a relatively light structure in such a way that the restoring force was reduced to a minimum. He ordered the construction of some rigid-sided models to begin.

Approval of the floating breakwater scheme was given by the Combined Chiefs of Staff at the Quebec and Washington meetings. They were worried by the possibility that the concrete caissons would be inadequate in scale to shelter Liberty and large stores ships and they were uncertain as to how effective fixed breakwaters would be. The alternative proposal of lines of floating ships proved, after experiments, to be impracticable.[5] The mooring requirements were astronomical, while the problem of anchoring broadside-on to an incoming sea presented what were considered to be unsurmountable difficulties.

Lochner was in the Admiralty team summoned to Quebec and while abroad he heard that the first experiments with a hard version of Lilo had been successful. These units were cylindrical in shape like the soft Lilo and were capable of reducing the height of waves by as much as 50 per cent. As soon as Lochner returned to London, the Admiralty ordered that full-scale tests with hard Lilo, now renamed Bombardon, should be put into operation. (The code name Bombardon was intended to suggest that the units were connected with anti-aircraft defences.) At the same time the Lilo trials were to go forward. Even if they were not used, useful lessons might be learned, particularly in regard to mooring.

By 15 October the trials of model Bombardons were com-

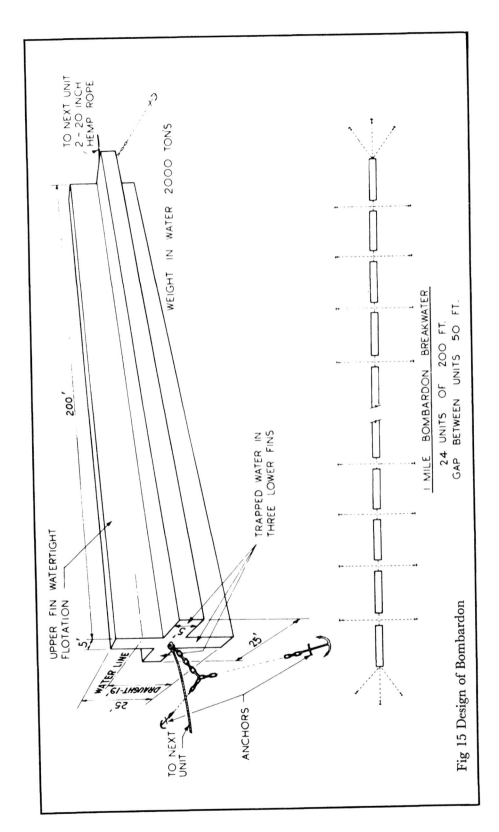

TO NEXT UNIT
2 - 20 INCH
HEMP ROPE

WEIGHT IN WATER 2000 TONS

UPPER FIN WATERTIGHT
FLOTATION

200'

5'

WATER LINE

DRAUGHT - 19'

25'

5'

25'

TRAPPED WATER IN
THREE LOWER FINS

TO NEXT UNIT

ANCHORS

1 MILE BOMBARDON BREAKWATER
24 UNITS OF 200 FT.
GAP BETWEEN UNITS 50 FT.

Fig 15 Design of Bombardon

64

pleted and the unit had now evolved to a hollow cruciform shape.[6] It was to be 200ft long, 25ft high, with a beam of 25ft. The top arm was enclosed, being composed of watertight compartments made from welded $\frac{1}{4}$in mild steel plate. The bottom and side arms were constructed from mild steel angles and plate in sections bolted together. The use of bolts in sea-going craft was highly unorthodox and had not been intended by Lochner, but was adopted because of the shortage of riveters and welders. The bottom and side arms were open so that when the unit was floating it was steadied by a mass of some 2,000 tons of water. The actual weight of steel used was, in fact, just under 300 tons. (See Fig 15.)

The Bombardon fulfilled the principles which Lochner had discovered on his trout pond that spring, ie, that a wavelength can be interrupted and reflected by a floating barrier, provided that each of the natural periods of that barrier is longer than the period of the longest wave to be stopped. (Like many inventions the principle of the Bombardon had already been put into practice. In a mid-nineteenth century painting of Brighton pier belonging to Iorys Hughes, a floating breakwater similar to the Bombardons is to be seen anchored off the beach to provide shelter for small craft.) The Bombardon was intended to break up seas and in a force 5 wind, or half a gale, it was hoped that it would reduce the height of waves from 8ft to about 3ft.

An initial order for seventy-five Bombardons was made but a change in layout later required 115 units. The gap between each Bombardon had to be large enough to absorb the relative motion but small enough to prevent waves passing through and reforming on the other side. Lochner and Penney, who did the mathematical calculations, decided on a gap of 50ft. Reduction of the waves was obtained by two, rather than by one, lines of Bombardons moored 800ft apart. The Bombardons were to take no more than a week to moor in position. In February 1944 the total number of units was reduced to ninety-three on account of the shortage of material. Even so, about 20,000 tons of steel were required.

The principal breakwater, however, would have to provide shelter for ships discharging supplies at the pierheads and for the barges, DUKWs and other craft ferrying cargo from ship to shore or to pierhead. While the naval staff maintained that blockships would solve the problem, civil engineers appointed by Tn5 believed that the answer lay in concrete caissons. In this case, however, the caissons would have to be constructed several hundred miles from the harbour site and would have to be towed across the Channel and sunk, possibly under enemy fire—a very different proposition from the orderly procedure of the port engineer.

Twenty-six years earlier, as it happened, Churchill had proposed the use of concrete caissons 'without expert assistance', as part of a plan to capture the Frisian islands of Borkum and Sylt.[7] He had envisaged the erection of an artificial port provided by caissons, towed across the North Sea and sunk in position in depths varying from 40 to 80ft, allowing room for sufficient freeboard, but no opportunity to put this scheme into practice ever arose. Concrete floating structures were, it has been seen, practicable, but no caisson had been built higher than 30ft. Conditions off the Normandy beaches would require a structure double that height.

Shortly after Conference Rattle Bruce White, on his own initiative, appointed what became known as the Artificial Harbours Committee. It met at Montagu House, a stone's throw from COHQ in Richmond Terrace. The members, all of whom had port engineering experience, were Reginald Gwyther, Colin R. White, J. D. C. E. Couper, Sir Leopold Saville, and occasionally an American engineer, Lt-Col L. B. Roberts. Lt-Col Ivor Bell of Tn5 acted as secretary. Gwyther became the moving spirit of the committee and was responsible for the preliminary design of the fixed breakwaters.[8] He had been awarded the Military Cross in World War I while serving in the Royal Engineers and in the following years he had designed bridges, docks and harbour extensions in Malaya and the United Kingdom. Recently as a partner of Coode, Mitchell & Vaughan-Lee (now Coode & Partners), he had been responsible for the design and construction of the Kut barrage bridges over the river Tigris.

The early meetings of the committee were considerably handicapped by lack of knowledge about the Overlord plan, what was required in the way of a harbour, and the extent to which, if any, the Bubble and Lilo breakwaters had been developed. However, the committee selected seven possible beaches where supplies could be discharged around the Cherbourg peninsula and the mouth of the Seine. They included the beaches north of Caen which were, in fact, the ones to be finally chosen. Also lacking was any navigational and metereological data on the area the committee had been told to investigate. When a naval officer from the Overlord planning staff attended one of their meetings, he merely produced some photographs of beaches, later withdrawing—with the photographs. The committee concluded he had been sent to find out what they were doing. Fortunately the information they required did arrive—just before the Quebec conference.

After that conference, on 24 September, Tn5 set up two civilian committees.[9] The first was called the Caisson Design Committee. It consisted of Gwyther and two eminent consulting engineers, W. T. (later Sir William) Halcrow, who had recently advised the Government on the construction of deep air-raid shelters and ordnance factories, and Sir Ralph Freeman, celebrated for his design of the Sydney harbour bridge. (Alarmed by so enormous a task and the lack of time in which to complete it, two members of the original Artificial Harbours Committee had resigned.) The second committee was called the Production, or Contractors' Committee.[10] It was headed by Sir Malcolm McAlpine, chairman of Robert McAlpine & Son, who had had great experience of dock and harbour construction. The other members were Storey Wilson and Norman Wates, chairman of Wates Ltd and responsible, as already noted, for constructing the prototype concrete floats. Wates had a wide experience of reinforced concrete work and possessed great organising ability. Both committees worked independently, liaison being provided by Colonel (later Brigadier) J. A. S. Rolfe, an experienced harbour engineer and deputy to Bruce White.

The designs had to conform to the following specifications. (1) They must be strong enough to withstand 8ft high waves about 120ft in length. (2) Their height should allow sinking in

50ft of water and they should have at least 6ft of freeboard at high tide. (3) They had to be capable of being sunk by opening valves while the water inside must not be allowed to upset their position. (4) They must be capable of being towed across the Channel at $4\frac{1}{2}$ knots. (5) They must be simple to build and not require labour and materials beyond the capacity of the Ministries of Labour and Supply.

None of the designs, which included American as well as British proposals, were considered to be entirely satisfactory. The main reason for this was the Contractors' Committee's doubts that some 140 units could be completed in the time available—six months. The committee was thus compelled to produce its own design, incorporating the best features of what they had seen. Called the Anglo-American design (see Fig 16), it was box-like in shape, with box ends, 200ft in length and 50ft in width and would displace some 6,000 tons. It was divided into twenty-two compartments, or cells, made of thin concrete and flooding valves were to be let into the cell walls where necessary. The sides, also of reinforced concrete, would be no more than 1ft thick. The lower part of the caisson, up to about 6–12ft above water level, would be constructed in a dry dock and floated out, the remaining part of the structure being built from a platform let in to the caisson. This would provide facilities for erecting scaffolding, incidentally avoiding the necessity for men not used to heights having to work on an unbroken surface rising up to 60ft. The platform also contained towing bollards and other equipment and was to be useful in other respects, as will be seen. Anything likely to slow down construction, including splays, fillets, hooks and bent rods, was eliminated.

This design when examined by the Design Committee was criticised, firstly, on the grounds of instability.[11] The upper part of the caisson was therefore thickened to 15in at the lower, and to 14in at the upper sections. Secondly, towing tests with models at the National Physical Laboratory, conducted by J. L. Kent and F. H. Todd of the Ship Division, showed that square-ended caissons would take twice as long to cross the Channel as caissons with swim-ends, ie, they would have to be towed at 4 knots as opposed to 2 knots. The swim-ends would,

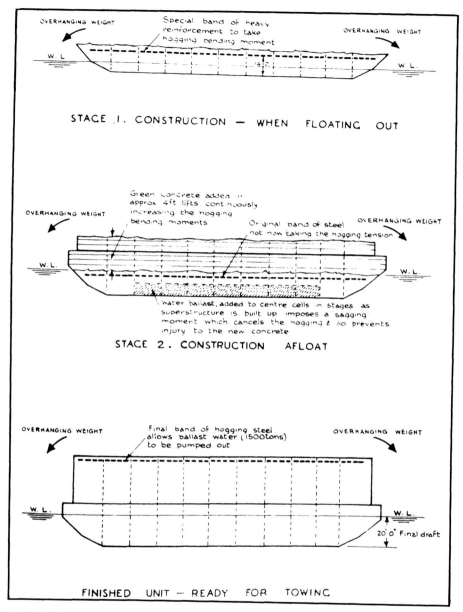

Fig 16 Construction of A1 concrete caisson

of course, produce scouring underneath them and this was why the Production Committee was adamantly opposed to their introduction. But as it was essential to cross the Channel during night time it was decided to ignore this factor. Construction was to be restricted to six types of caisson. On account of the varying depth of the sea bed, the caissons had to be made in several dimensions.

TABLE 1

Dimensions of all Classes of Caisson

Unit		Height	Length		Breadth at water level		Displacement	Draught	
ft	in	ft	ft	in	ft	in	tons	ft	in
A1		60	204		56	3	6,044	20	3
A2		50	204		56	3	4,773	16	4
B1		40	203	6	44		3,275	14	0
B2		35	203	6	44		2,861	12	5
C1		30	203	6	32		2,420	14	3
D1		25	174	3	27	9	1,672	13	0

Initially, 147 caissons were required for the two Mulberries. At a late stage in the planning of Overlord, however, it was decided to extend the life of the harbours from ninety days to the end of the year at the earliest. This programme of what was called 'winterisation' demanded more caissons which would 'double-bank' or reinforce the existing breakwaters. The building of caissons therefore continued throughout the summer of 1944 raising the number of caissons built to 212.

The design of the caissons had to be modified considerably because of the sites on which they were built. As will be explained in Chapter 5, this was mainly due to the shortage of dry docks. Nearly three-quarters of the caissons had to be built in two stages, in the manner described above.

This procedure gave rise to several major engineering problems.[12] The original design had assumed that the structure would not have to take up the strains and stresses of floatation until completed. These strains and stresses were distributed throughout the structure and if it was incomplete the lower part of the structure would be overloaded. Moreover, when only partly constructed, the structure would be of less weight and would therefore float higher in the water; so much so, that the swim ends would be unsupported. Since these were necessarily of a more solid construction than the central section there would be a tendency for them to sag on either side of the centre which, being in the water, would be the main area of support.

Oscar Faber, who had already been engaged in the design of the Lilo keel, drew attention, with several of the construction engineers, to this difficulty before work on the caisson began. The changes involved strengthening the sides 14ft above the bottom with additional steel reinforcement. The swim ends also had to be lightened by reducing the thickness of the concrete. This redesign meant that the orders for steel reinforcement which had already been placed had greatly to be enlarged, causing considerable confusion.

The second major engineering problem arose from the fact that the caissons were one homogeneous whole throughout which the strains and stresses were spread. The process of building up the structure after the shallow keels had been floated out of the basins was carried out in stages of about 10ft at a time. As each stage was added to the existing structure it became a part of the whole, and this had to bear the floatation strains and stresses which automatically spread to it. Each new stage, when added, was of green concrete which, until it was set, was unable to take strains and stresses; had it been allowed to do so, it would have cracked and spoilt the whole job. It was necessary therefore to devise some means whereby each new lift of concrete could be protected from these strains and stresses until it was set and strong enough to take them. The two ways of solving the problem were, firstly, by building up the central portion sufficiently in advance of the ends, or, secondly, by adding kentledge, or ballast, to the centre of the unit as the concrete was brought up. The former method was favoured by the contractors, but it was much slower than taking each lift of concrete through from end to end, and it was calculated that it increased the time taken to construct each unit by about three weeks.

The ballast considered for the alternative method was in the form of concrete blocks, or of shingle, but both these were impractical with the mass of scaffolding filling up the interior of the caisson and were difficult to assess correctly for weight and disposition. Instead, one of the contractor's engineers suggested that water should be used; it was simple for the contractor, it lent itself to exact calculation and it entailed very little loss of time.

As if these changes were not enough, a further addition to the design was imposed after sinking trials at the NPL at Teddington gave rise to the fear that the caissons might heel over after the valves had been opened for flooding.[13] The NPL recommended that stability should be increased by adding 1,024 tons, or concrete ballast, or by building 5ft-high baffle walls to reduce the free water surface until the caisson had developed sufficient extra stability by admission of water up to the top of the baffle walls.

Tn5 ordered the addition of 5ft walls supported by an additional layer of concrete. By this time work was already advanced on the sites and the howl of protest that went up 'could have almost been heard in Berlin.'[14] 'We were forced to compromise', wrote W. J. Hodge from the Port of London Authority, then a major in charge of the Tn5 caisson design team, 'with a 15½in layer of concrete ballast plus some small baffle walls of 4½in thick brickwork only 18in high'. In the event the fears of being unable to sink the unit on an even keel proved to be unsubstantiated. There was only a slight list caused by the platform on one side becoming awash, but the list was corrected a few minutes later.

Finally, when the Civil Engineer-in-Chief's department in the Admiralty was drawn into the Mulberry project, fears arose that the caisson was not strong enough to take a pounding from high waves and might even be shifted when resting on the sea bed. The problem was given to Penney to solve.[15] He discounted the possibility of the caisson moving; on the contrary, he thought that if the sea bed was of shingle they would dig themselves into it. Nor was he worried by external wave pressure. What *did* concern him was the possibility of severe internal pressure being created by water being unable to escape through the lower set of valves (a possibility which also occurred to Col S. K. Gilbert, deputy-commander of the British Mulberry construction force). Penney proved in the event to have been remarkably percipient. If there had been more time, the answer would have been to roof over the caissons as was, in fact, done with some of the units built after D-Day.

As a result of the Admiralty criticisms, some strengthening to the steel reinforcement was introduced in the last units to be built.[16]

4 Blockships

When General Dwight Eisenhower was appointed Supreme Allied Commander and General Montgomery commander of the Anglo–US assault force, they reviewed the plan for the assault and concluded that an attack of five rather than three divisions should be made. This meant that the beachhead would have to be extended to about fifty miles of coastline, so that three additional beaches, apart from the Mulberries, were now required.

The Admiralty pointed out that in the event of bad weather up to 4,000 small craft would need shelter.[1] The Mulberries could not provide the space for these craft in addition to the vessels normally using them; moreover, the Mulberries were situated comparatively close together in the centre of the invasion coastline. Should a gale suddenly develop all the landing craft would be unable to get back from the flanks to the harbours.

The solution to this problem would be the provision of five small landing-craft shelters (Gooseberries), two of which would be embodied in the layout of the Mulberries. In the case of Mulberry B the blockships would help to close the gap between the caissons and the Calvados reef, a natural barrier against the weather from the north-east. Although giving some slight shelter to a small portion of beach, their primary object was to provide a breakwater behind which landing craft could shelter in heavy weather. (See Fig 17.) Discharge of stores would cease while the craft were sheltering, but could start again directly the weather moderated. As the breakwaters were for the protection of shallow-draught vessels, blockships were suitable for the purpose. At the same time it was only by the use of self-propelled blockships that breakwaters could be established by D-Day plus three. Shelter had to be provided from the start.

Tn5, as already seen, was strongly anti-blockship. In war, Gwyther maintained, blockships had usually been used to obstruct shipping rather than to protect it.[2] 'When sunk', he wrote, 'they rarely finished up where they were wanted ... Owing to the small area available they jeopardised the whole

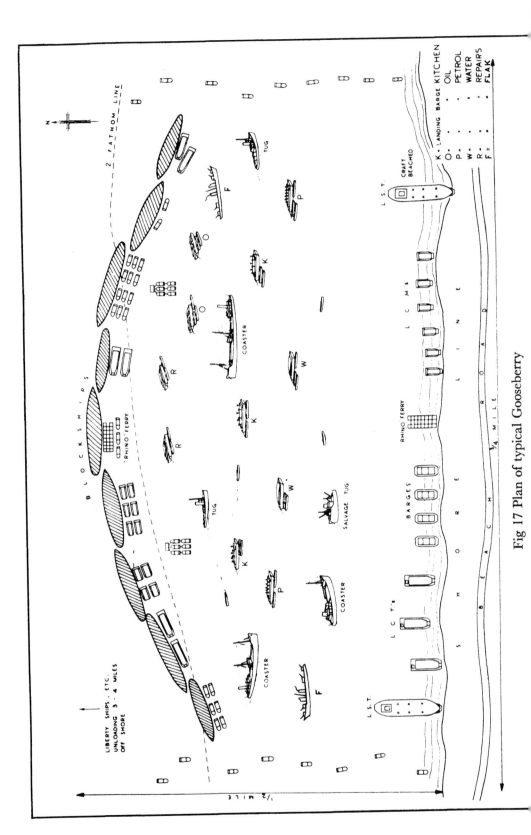

Fig 17 Plan of typical Gooseberry

74

scheme.' He quoted the comparative failure of the British attempt to block the harbour entrances at Ostend and Zeebrugge when the navy failed to sink the ships in line.

Nevertheless, when the matter was discussed at a Chiefs of Staff meeting presided over by the Prime Minister on 24 January 1944, the First Sea Lord, Admiral Sir Andrew Cunningham argued powerfully for the provision of blockships.[3] It had after all been agreed at Quebec that they were suitable for forming breakwaters in shallow water. About seventy unserviceable merchantmen must be found and, if there were not enough of them, some old warships would have to be scuttled. While the engineers might be sceptical about blockships, the Admiralty were equally doubtful about caissons and the time they would take to be towed across the Channel and sunk in position.

Considering the acute shortage of shipping, the use of blockships was a daring proposal and there was a good deal of reluctance in accepting it, particularly on the part of the Prime Minister. But the matter was eventually clinched by the Chief of Air Staff, Air Chief Marshal Sir Charles Portal, who remarked that it would be absurd to risk the assault for the sake of a few derelict ships.

The planners soon realised that the Gooseberries would not only provide shelter but could form an administrative centre with facilities for first aid, repairs, fuelling etc, while the superstructures would provide accommodation for the crews of the assault craft. Altogether 25,000ft of breakwater would be provided by blockships, of which the Americans agreed to provide twenty-two.

As soon as the Chiefs of Staff had decided to sacrifice the shipping, suitable vessels were sought In order to conform to the high rise and fall of the tide, the depth at low water and the freeboard at high water could not, together, exceed about 18 to 25ft. This meant that the ships could only be used in about 12 to 15ft of water; that depth was adequate for all landing craft but only sufficient for the smallest types of coaster. The blockships had to be around thirty years old, by which time they were usually often in the repairer's yards for maintenance.

The task of preparing shipping as blockships was the responsibility of the Ministry of War Transport.[4] Conversion took place at Rosyth on the east and Methil and Oban on the west coast of Scotland early in May. Stringent security was observed and after conversion the crews were not allowed to go ashore. The operation itself was fairly simple. The ships were ballasted to draw about 18ft of water and this had to be done with some precision in order that the ship should sink on an even keel. Over a quarter of a million tons of ballast were used. Next, 10lb of amanol explosive charges were fitted to either side of each hold about 3ft below the waterline and were connected by leads to an electric firing key on the bridge. (A problem here was that rats ate the insulation of the cables!) Ten charges were placed in the larger and eight in the remaining ships.

The Royal Dockyards prepared four warships selected to join the merchantmen. They were the 32-year old British battleship *Centurion*, the French battleship *Courbet* of the same age, the old British cruiser *Durban* and the stripped hulk of the Dutch cruiser *Sumatra*.

Two weeks before D-Day all the blockships assembled off the west coast of Scotland preparatory to sailing south.

5 Construction and Assembly of Components

The mass construction of the components of Mulberry did not begin until December 1943, so that there was less than six months in which to complete the basic requirements for the landings. It was calculated that a labour force of 30,000 men would be required immediately. Fortunately the construction of camps, runways and other installations connected with the mounting of Overlord had by now largely been completed so that the organisation evolved by the Ministry of Works to keep track of allocations of men for particular jobs could now be adapted for the Mulberry programme. Recruits were encouraged to move to areas like London by the offer of accommodation, travelling allowances and other inducements. Ernest Bevin, Minister of Labour, took a close personal interest in the scheme, smoothing out problems with the unions and ensuring that the varying skills of the work force were used to the best advantage. Within sixteen weeks over 25,000 men had been transferred to build Mulberry components. When construction reached its peak a total of 45,000 men were being employed.[1]

The construction of the Whale pierheads and floating roadways required a prodigious amount of material. By January 1944 the Chiefs of Staff had increased the number of pierheads to twenty-three, eight of which were to be spares. Auxiliary equipment included fourteen buffer pontoons and twenty-four intermediate pontoons. The total requirement for Whale demanded 60,000 tons of steel. Interchangeable parts amounting to 200 had to be prefabricated. The floating roadways were increased from six to ten miles, of which three miles were to be spare. This total included 120 80ft spans and eight shore ramp floats. The roadways were carried by some 670 floats, of which 470 were to be built in concrete. Materials consisted of 30,000 tons of steel and concrete. Prefabrication of the pierheads was divided among 300 firms, and of the roadways among 250 firms. The total work force employed amounted to 15,000.

As soon as Tn5 appreciated that most of the floats would

have to be made of concrete rather than of steel, it issued instructions that the design should allow sections to be prefabricated as far as possible, then assembled, cast together and launched at a suitable riverside slipway. Wates, responsible for producing the prototype floats, were asked to carry out their final concreting and assembly in the Southampton area. Local firms, like Marley's Tiles, made the precast slabs. However, when the unanticipated production of the Bombardons made further inroads on the limited supplies of steel, the number of concrete floats had to be increased. Additional firms were now required both for assembly and for making precast slabs. This decision was made late in January 1944 and the target date was set for 31 March.

Many of the firms called upon to make the slabs were more used to making fence posts, and those not always of the highest quality. Victor Wigmore, a well-known authority on concrete, and consultant to Mouchel's, became adviser to the Ministry of Supply and to see that the work was done properly he had to visit some eleven firms in the London area, in addition to the work in progress on the south coast.[2] In the former area such firms as Orlit's, Girling, W. & C. French, Costain, Stewart's Granolithic Co and the Trussed Concrete Steel Co were all involved.

The concrete had to be of first-class quality. No waterproofing compounds were added, though an admixture was used as an accelerator for in-situ concrete during the severest part of the winter. All the aggregates had to be tested before being accepted and in areas where the sand was not good enough, an alternative supply had to be transported by truck, often from some distance away. The mix for the concrete was 11cwt cement, $13\frac{1}{2}$cu ft of sand and 27cu ft of $\frac{3}{8}$in aggregate. The moulds were then vibrated from $1\frac{1}{2}$ to 3min. As already described, the outer panels were no more than $1\frac{1}{4}$in thick, replacing the 2in-thick slabs of the prototypes; they were 2ft 6in wide with steel reinforcing and made in lengths to suit their position in the float.

The panels were 'cured' with water for the first three days and were usually transportable after the fourth day. All the edges of the slabs were treated with a water spray jet when only a few hours old in order to ensure a proper jointing surface. Wigmore constantly found faults: the moulds were often distorted and had to be scrapped, the concrete was either too dry or too wet, or the vibrating was weak. When ready to go to the assembly area, the panels were stacked vertically, or even horizontally, in lorries.

No delay could be tolerated throughout the entire process. On one occasion a supply of reinforcing bars for the panels arrived late. This meant that on the following day concreting would have to be delayed as the bar fixers could not get the bars into position in the time available. In order to keep to schedule, additional labour was brought in, including one of the directors of the firm and the drawing-office staff who worked throughout the night placing and tying reinforcement bars. By the following morning the job was completed. The precast work was finally concluded at the end of April, a month behind schedule though corresponding to Wigmore's more realistic appreciation.

Attempts were made to lighten the concrete and yet maintain its strength, in particular in the floats near the shore end of the piers. Wigmore proposed the use of foamed slag as the fine aggregate and normal shingle for the coarse aggregate. 'There was horror', wrote Wigmore later, 'at the idea of using a light-weight aggregate for a water-retaining structure.' Tests were made up to 150lb psi water pressure on a $1\frac{1}{4}$in-thick slab. This was held for six hours with no sign of percolation. Seven years later Wigmore examined a concrete float on the beach near Bognor Regis. The aggregate looked like granite, but under a glass it proved to be foamed slag. The concrete was in excellent condition and the bars uncorroded.

Marchwood, now the site of Southampton power station, (see Fig 18), was the main assembly and in-situ concreting area for the floats. Known as No 1 Port and Inland Water Transport Repair Depot, it had easy road and rail access. Wates prepared a slipway and other installations.[3] The water frontage was limited but adequate for the purpose and part of the area was roofed over so that work on the floats could continue in bad weather. Work, in fact, continued despite air-raid warnings, raids, and later flying bombs.

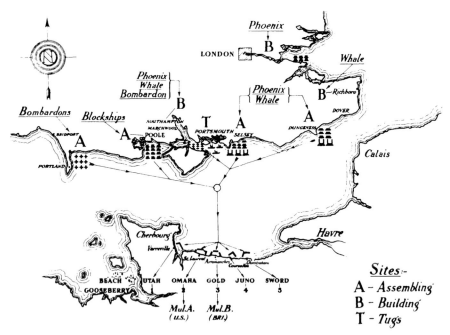

Fig 18 Assembly of the Mulberry components

Two other assembly sites were chosen—the promenade at Southsea and several basins which had been excavated at Beaulieu, where naval ships had been built since Tudor times. The work at the Southsea promenade was carried out by John Laing for Wates and as the site was surrounded by hoardings, few realised the significance of what was going on behind them.[4] Floats were also assembled in the West India, Erith and Russia docks in London, this work being done by John Mowlem and Melville Dundas & Whitson, firms which had just completed building a batch of caissons.

Constant supervision of the in-situ concreting was essential. One-ton cranes lifted the panels and tubular and timber scaffolding held them in position until the concreting was complete. Assembly and completion of concreting was brought up to a speed of four days per craft. Five days were specified for maturing the last-placed concrete, though floats were sometimes launched in less time.

The next problem was to move the heavy craft to the water's edge. The method varied according to the site. At Marchwood a special carriage was run under each float on the building berth. The float was then rolled down the concrete slipway into the water and floated off the undercarriage. At Beaulieu, all that had to be done was to let water into the basins and float the Beetles out. At Southsea, where there was a gently sloping beach, the Beetles were launched down slipways into the sea. In the London dock area, the quay was 8ft above water level, so that the floats had to be wheeled on bogies to an overhead gantry and then slung into the water. On one occasion a 46-ton Beetle was dropped 4ft onto a concrete bed. The damage was only slight and was repaired in a few hours.

The labour force was recruited from the most unlikely quarters and included pimps and flower-sellers. The more skilled were trained to check the craft for leaks after launching and a quick-sealing fixative called Sika was used, as repairs had to be carried out against an outside head of water. At this stage very little water seeped through, but when the craft were towed in the seaway they were subjected to more severe strains, as will be seen. The final fitting operations took place on the water.

The construction of the spans was simplified by the design of jigs and templates, enabling sections to be welded together by numbers (this had been foreseen by the Ministry of Supply). The template was designed by Lamond of Braithwaites. Being interchangeable, it enabled a number of other firms to manufacture plates and girders. Many thousand tons of steel were saved by welding wherever practicable.

Wernher, in his rôle of co-ordinator, mobilised rail wagons, of which there was a scarcity, to bring the supplies from the steel mills to the factories. Colonel (later Brigadier) Ian Mackillop, a fine staff officer, organised these movements. He later became Deputy Quartermaster General 21st Army Group. Travelling all over the country, Colonel T. B. Bunting kept Wernher in touch with the state of progress and his first job was to make a rapid survey of existing resources during the Quebec conference.[5]

By the end of December the members for the spans began to arrive by road and rail at Marchwood and at Richborough on the Kent coast. By early March 1944 the assembly area was ready. The work was to be done by depôt troops, using both special and rat-tail spanners. They were good tradesmen but many were category men unfit for service overseas. However, with careful coaching, they proved to be more than adequate for the job and as each span weighed between 28 and 36 tons it was no easy task.

A rail for a 5-ton crane ran alongside a building track and on the other side of the crane track there was a storage area holding materials for two spans. As assembly of the spans went ahead, materials could be unloaded for the next day's work. A span assembly party consisted of twelve men, including one steel erector. If all went well they could erect one span in a day.

The sequence of assembly went as follows.[6] The two main girders were assembled flat on baulks of timber. On completion they were raised into an upright position by crane with about 15ft in between. The components connecting the main girders were now hoisted into position and the cross girders were then fastened, starting with the centre girder. Next the steel decking was laid. After completion the span was jacked up and the transverse timbers removed. The span was then lowered onto trollies and towed to a storage bay where it was checked for defects. The spans were finally moved to the jetty, where they were lifted by a 30-ton crane and placed on the Beetles tied alongside.

When the spans had been fixed to the steel transoms on the floats, they were made up into tows, each of six spans, the tow being 450ft long. Each tow was equipped with tool-boxes and stores such as saddles, wood packs, pivot plates, etc and was then ready to be towed to the assembly area in the Solent. By D-Day the troops at Marchwood had assembled about three miles of Whale.

Richborough had been the base for the Romans when they occupied Britain in AD 43—the last but one successful cross-Channel invasion. Here the steel floats were erected and spans placed on them.[7] The all-welded float units, after being tested for leaks, were transported from the manufacturers by road and rail to Richborough. They were bolted together by a force of about 1,000 sappers assisted by 300 Americans from 108 Construction Battalion. It was appropriate that this force produced the first item of Mulberry equipment (15 spans) to be handed over to the Americans.

On completion floats were parked until required in nearby Pegwell Bay. When the tide went out floats often stuck in the mud and had to be released by hosing their bottoms; otherwise they would have been submerged at high water. As at Marchwood, about three miles of pier (228 steel floats) were completed and by 27 May the minimum operational requirements in both cases had been achieved. In the end the sappers were completing twelve spans a day.

The pierheads were to be built at three sites in Scotland and Wales and, as with the other components, new building sites had first to be completed. This preparatory work took up to ten weeks and the units now had to be ready in four weeks instead of the four months originally anticipated.

Welding was to be used whenever possible to save steel and much of this work was done by men from trades such as hairdressing and tailoring who had had no previous knowledge of welding techniques. Small parties of trained men were sent by the Ministry of Supply to give instruction on the site and the design of the units were simplified to assist the inexperienced work force.[8] These plans were prepared by George Maltby of Messrs Redpath, Brown & Co who also organised a steel construction firm to turn out the steel plate for the pontoons. Findlay's were responsible for both yard and pontoon construction at Leith, where it was decided to build thirteen pierheads and sixteen intermediate pontoons. They were also responsible for the more complicated key work carried out at Motherwell.

The site chosen at Leith was a partially-built extension to the docks abandoned on the outbreak of war.[9] George M. Carter Ltd, a firm of steel erectors, put the yard into working order and, under the direction of R. Newson, they built and launched the pontoons. The welding was undertaken by the Lanarkshire Welding Co under Robert Macdonald. The total work force came to 600 men, 200 of whom were welders. After launching, the hull was towed to an adjacent fitting-out basin. Here the pipe work was carried out by George Martin, master plumber of Motherwell; electrical fittings by John Robertson of Glasgow; and joinery by Messrs J. & R. Watson of Edinburgh. The first pierhead was launched on 26 January 1944.

Three pierheads were built by Port Construction and Repair Companies on a slipway to the north of No 2 Military Port at Cairn Ryan, where there had to be a change-over of men after the building of the first two craft as training for Overlord had now started on the English south coast. Many of the sappers were unable to weld and instruction took place whenever the men were free from other duties. The building and launching of a fourth pierhead was done mainly by civilians sent over from Leith. Capt Witcomb R E then took over the equipment.

Five pierheads and four buffer pontoons were built at Conway's Morfa estate, not far from where the Hippos had been launched. Messrs Joseph Parks & Son of Northwich and Waring & Co of London were the building contractors, while Holloway's, under the direction of Harold West, were responsible for traversing and launching the craft. Oleg Kerensky, son of Alexander Kerensky, leading minister of the provisional Russian government in 1917, was in charge of the buffer pontoons.[10] He had escaped to England with his father, subsequently embarking on a successful career as an engineer. At this time, however, he had not become a British subject and was required to conform to the regulations affecting an alien—and this in spite of being associated with the top secret Mulberry! Kerensky was an admirable leader of men, resolving labour disputes (the workers were unaware that he was 'white' rather than 'red') and ensuring that the canteen was being run properly, in addition to his technical responsibilities.

Like the caissons, the pontoons were built one behind the other and then launched sideways. A force of some 900 men swarmed into Conway where they were housed in hostels, holiday boarding-houses and several convalescent homes. Work went on round the clock, ignoring the black-out. The welders operated under strong arc lights which heightened the contrast of the surrounding darkness. Many men, including Kerensky, developed eye strain, known locally as 'arc eyes', and had to have medical attention. Often men fell asleep where they were working. Other problems arose such as the occasion when a particular welding job required respirators. Cotton-wool pads were immediately improvised by the Conway Women's Voluntary Service. Here, as elsewhere, the lack of skilled labour was evident. Carpenters were taught to mark steel and other jobs were duplicated by workers who had no idea of the purpose of the strange craft they were building beyond the fact that it was 'something big'. Launching the craft took place with the minimum of ceremony, a bottle of beer usually being broken over the steelwork.

After being tested for leaks and fitted with towing gear, the pierheads were towed by night to Southampton where the spuds were to be installed and fitted with last-minute protective equipment against air attack.[11] Dorman Long made the heavy spud columns. The work took place near the King George V dry dock which had a 50-ton crane suitable for lifting the spuds. Sappers from Marchwood assembled and riveted sections of the spud columns and welded on the feet. The spud cables were reeved and the spuds stepped into position with the help of the crane. The steel cables were protected with armour plating where these were exposed over the top pulley of the spud. This was provided by Brightside Foundry & Engineering Co. The control cabin was also armour-plated and Oerlikon anti-aircraft guns were installed. As no instructions were provided, Witcomb had somehow to teach the crews to handle the guns, which were later withdrawn after an over-enthusiastic crew fired a shell through one of HMS *Rodney's* smoke stacks at Mulberry B. Various adaptations were made to suit the particular operational functions of each pontoon. Some were fitted with hinge bearings for buffer pontoons and had to have internal stiffening. Others were fitted with Baker fenders involving the manufacture by John Elwell of 5,400 steel brackets on which to hang them.

It was not surprising that the contractors failed to keep to schedule. The first batch of pierheads should have been ready at the end of March. R. D. Davis, Deputy Director-General RE Equipment, in the Ministry of Supply, was responsible for organising the manufacture of machinery and equipment for the spud pontoons. Appreciating the anxiety of the Chiefs of Staff and the Overlord planners, he addressed a meeting of the British Structural Iron Association in London during the first week of April.[12] The response was immediate. A force of 300 welders from various firms was rapidly assembled under the direction of Dorman Long at Southampton. They had no knowledge of the work they had to do except that it was vital and desperately urgent. The first five pierheads were completed by 10 May and, according to Davis, 'saved the situation'.

Work was also impeded by the lack of mobile cranes. On investigation, it was discovered that a number were being used for shipping coal in South Wales. They were immediately commandeered by Wernher and brought to Southampton. So, too, were sixty petrol-driven welding sets and lorries.

When each pontoon was completed in detail it was tested and alternately crewed with British (969 Port Construction and Repair Co) or Americans from 108 Construction Battalion. The Americans were usually in the charge of chief petty officers, many of whom had been oil drillers in Texas and were knowledgeable and highly competent in their work. The pontoons were then towed away to anchorages in the Solent or off Selsey. Three pontoons were fitted out at Woolwich and Falmouth. The story is told of the naval look-outs at Dover who reported a strange-looking object in the distance undulating through the water like a sea serpent. It proved to be a bridge tow low in the water, the spans sticking above the horizon, en route to Selsey.

By mid-May fourteen pierheads were on tow or had arrived at Southampton, leaving one outstanding for the initial requirement. All the pierheads bar one were completed before the end of June, though four of them still had to be fitted with spuds.[13]

The late arrival of the pierheads was particularly resented by the Americans, probably unaware of the constructional difficulties but anxious to do some training before crossing the Channel. In spite of repeated requests they did not receive one until a few days before D-Day. According to Capt Clark, in command of Force 128, however, Admiral Tennant was unperturbed about the delay.[14]

Perhaps more serious than these delays was the discovery of a defect in the pierheads.[15] When beaching their LSTs on the buffer pontoon, crews found they could not open the bow doors because the incline was too steep. There were two ways of solving this problem; either by raising the slope so that the craft were high enough to open their doors, or modifying the bow doors themselves and approaching with them open. The Admiralty Corps of Naval Constructors favoured the former alternative and proposed building a concrete tumulus for the purpose.[16] But Pavry did not like the idea because a vessel could easily foul the obstacle whether it approached with bow doors open or shut, and particularly if it was not dead on centre.

Pavry expressed his misgivings to Hickling, who was fretting because he had been kept in the dark about the operation of the pierhead. In no time Pavry was invited to meet the British LST flotilla commander, Lt-Cdr Hore-Lacey, RN. The latter likewise was in no doubt of the unsuitability of the Admiralty's proposal. The two officers promptly set off in a rowing boat and marked off with chalk the corners of the LST bow doors which were to be trimmed by oxy-acetylene welders. Hore-Lacey straightaway set about modifying his vessels so that should the Admiralty object, he would be able to reply that the change had already been made. The Americans similarly modified their LSTs.

Another defect discovered at this time was that tracked vehicles were unable to climb the steep steel incline of the buffer pontoon because of the slippery surface. Special wooden mats were then improvised so that armoured vehicles could get a grip. All these modifications to the buffer pontoon had, of course, to be completed at high speed in order to be ready for towing across the Channel.

The great size of the intermediate pontoons involved construction methods similar to the caissons. Twelve of these units were built by Wates at Marchwood and Beaulieu in sites chosen by C. D. Mitchell, who also supervised the work on the Beetles for that firm.[17] When the wide decks had been concreted, four days elapsed before the pontoons were transferred to the launching cradles. Fortunately that winter was a mild one and favoured work out of doors. At Beaulieu, Mitchell converted oyster beds into dry docks by cleaning out and then concreting the bottoms. The dock was then sealed off by timber gates constructed within a steel sheet pile coffer dam. When the gates were completed, the piles were burnt off at ground level when the tide was out. On completion of the pontoon the gates were removed by derrick crane and the unit floated out on a suitable tide. One unusual problem encountered by the builders of the lock gates was that crabs took a fancy to the felt covering on the face of the gates.

At Marchwood, the pontoons were built on wedged sills, or dwarf walls, from which they were transferred to greased slipways and pulled by two 8-ton winches onto cradles on the inclined launching ways. Because of their great weight they were sometimes difficult to start, but once in motion very little effort was required. Inspection after launching usually revealed very little leakage that was not easy to repair.

One incident at Marchwood temporarily alarmed Wigmore, who wrote:[18]

The test cubes crushed at seven days gave a result nearer 600 psi instead of the usual 5,000 psi (approximate). Panic stations. I went down straight away suspecting sabotage (lots of Irish labourers) and wondering whether something like sugar could have been put into the mixer. But after a check, I was satisfied as far as the aggregates and batching were concerned. There remained only the cement—the last thing I would suspect normally. Rapid hardening cement from a well-known firm was usually used. Not that it was unknown for bags to be interchanged. When I got back to the office I phoned the Chief Technical Officeer at the cement company and told him to come round to my office at once. He did so. I told him the story and asked him to investigate from the cement angle. I explained the urgency (it was already April). The worrying point was that this weak concrete joined the precast panels in the pontoon. To cut it out would have damaged the precast panels. There was no time to get new panels as the moulds had all been dismantled: the contract was finished. In two hours I knew the answer. What had happened was that the cement had come from a different works. The chemist had had a 'brainstorm'. The cement which had been produced was low-heat Portland cement but he had allowed it to go into rapid hardening bags!

Now I knew the truth I was able to get down to details of what was to be done. Fortunately I always had a larger number of test cubes made than were really needed, so in this case I should be able to check any estimate I might make. If this low heat cement was true to form I reckoned that in six weeks the strength of the concrete would be sufficient to give a safety factor of about two. Would that be sufficient? The consulting engineers decided to go ahead, especially when I said that the conditions in which the concrete would mature would ensure a good increase in strength with time. Everyone breathed again!

North of the Thames four pontoons were built by A. Monk & Co at Barking in close proximity to the Phoenixes. Two others were built by the Trussed Concrete Steel Co at Rainham. The idea that concrete could float was usually treated with scepticism and when, some time before the manufacture of craft for Mulberry, the first petrol barge of concrete was lifted by the Port of London floating crane, everyone stopped work expecting to see it sink.

The Phoenixes required a greater quantity of labour and materials than any of the other components and a small branch was set up in one of the directorates of the Ministry of Supply to co-ordinate construction. J. W. (later Sir John) Gibson was made Deputy Director-General of Civil Engineering (Special). Gibson, a civil engineer from Pauling's, had already held two important wartime appointments connected with ordnance factories and building construction and has been described as a 'forceful go-getter'. Under him were two directors, also engineers, one technical and one administrative, the latter being H. W. Phillips whose competence won great respect. For the purpose of administration the country was divided into three areas, North Thames, South Thames (these being the areas in which most of the caissons were built) and the Southern Area. In charge of the drawing office was W. V. Fuller who, according to a member of Tn5 in close touch with this branch, was able 'to lay his hands on a thousand of anything, tons, feet or pieces—within a couple of hours of a mere phoned request!' [19]

The construction of the caissons was allocated to seven firms, or groups, of consulting engineers. They were Oscar Faber, Sir Alexander Gibb & Partners, Sir Cyril Kirkpatrick, Rendell, Palmer & Tritton, a group of joint engineers which included Coode, Mitchell, Vaughan-Lee and Gwyther, W. T. Halcrow & Partners and Wolfe Barry, Robert White & Partners.

The pre-D-Day programme absorbed at its peak in mid-March 1944 over 22,000 men, many of whom came over from Ireland, thus involving an additional security problem. They included 10,500 labourers, 1,300 scaffolders, 5,000 carpenters and 770 steel fixers. Even this total was not enough. Bevin therefore arranged for 1,000 carpenters to be transferred from the Admiralty, while over 1,000 tradesmen were released from the army to build the caissons.

Men were chosen who were accustomed to working at heights (although as explained earlier, the gangway on the caisson provided a base for scaffolding). If all went well, it was estimated that one caisson could be built in about four months. Upwards of 132 would be required by D-Day.

Next, there was the problem of finding suitable construction sites. Ideally, the units would have all been built in dry docks. But the Admiralty had anticipated that dry docks would be subject to the requirements of shipping. They had made alternative proposals which were, in the main, those adopted by the Ministry of Supply.

The Admiralty selected the large King George V dry dock at Southampton for the assembly of the Bombardons. To the chagrin of the War Office, this dock had the space in which to build eight or nine caissons at a time. Instead, the Ministry of Supply was given the dry or graving docks at Tilbury and the graving docks at Goole and Middlesbrough.

XII Construction of Phoenix. Note the cylindrical 10-ton concrete clump anchor for use in emergency

A rapid survey of other dock facilities had to be made. The East India Dock and the South Dock of the Surrey Commercial Dock were capacious and about eighteen A1 caissons could be built in one stage in two batches, thereby saving time.[20] These docks were now drained of water for the first time since they had been built over 100 years ago. The possibility of dock walls collapsing due to water in the soil after the dock had been drained was foreseen and well points were installed behind the west wall. Nevertheless, on 8 December, three days after the dock had been pumped dry, the west wall collapsed and nearly 300ft of the south quay collapsed on 18 January. The building programme was not interrupted, although work on two caissons had to be moved to alternative sites.

In both docks hardcore was laid down on the casting site and, to protect men and equipment on the dock floor and minimise risk of damage from air attack, a concrete dam was built across the dock entrance. This had to be demolished with explosive charges upon completion of the units. At the South Dock, however, a coffer dam of steel piles was built. When the caissons were ready to come out, the piles were withdrawn and the units towed out by way of Greenland Dock. The clearance on either side of the exit was no more than $9\frac{1}{2}$in.

The second method was to build the caissons on a beach overlooking the sea and launch them down a slipway. Sites were chosen at Stokes Bay, near Portsmouth, and at Stone Point and Langston in the vicinity of the Solent. Thirteen caissons were launched at Stokes Bay by Messrs Holloway, putting to good use what they had learnt in Conway when launching the Hughes caisson. There were other problems with which to contend. The tidal range at Stokes Bay was only 13ft at spring tides, which meant that construction of the lower end of the slipway was only possible at big spring tides, occurring once a month.

The units were built in echelon, one behind another. Each unit was built on keel block walls at a height of about 20ft and then transferred to eighteen ball carriages using special wedges and clamps. The caisson was then hauled forward by winch and tackle until directly over a track parallel to the foreshore. It was then hauled along until directly over the launchway. The weight of the unit was now transferred from the ball carriages to the slipways and launched. The final stages of construction were completed while the caisson lay moored off the beach.

The caisson builders had unknowingly intruded on the testing ground of the swimming tanks, an element of the specialist 79th Armoured Division which was to be the spearhead of the assault. This was typical of the demands of space that arose during the preparations for Overlord in the final phase. Consequently the amphibians had to move to a more secluded beach.

The third method of assembly had never been tried before. A temporary basin in a suitable site was to be dug out and the units floated out after the first stage of construction had been completed. As the Production Committee warned, there could be 'a certain degree of unreliability [in] subsoil close to tidal waters.'[21] The Thames estuary was considered a suitable area and eight sites were chosen at Erith, Barking and Grays, where seventeen basins were excavated behind the river embankment. This work was carried out by a variety of methods including the use of drag lines, scrapers and bulldozers, the last-named being the most successful on account of the waterlogged nature of the ground.

Thirteen of the basins were large enough to hold two units at a time and on those sites which had more than one basin, the caissons were arranged side by side, the embankment being breached for each basin. On one site, however, three basins to hold five units were excavated, one behind the other, the innermost basin being nearly 1,000ft from the river bank.

Four more basins were excavated on a site between Russia Dock and Lady Dock in the Surrey Commercial Docks. The site was laid on the foundations of warehouses destroyed in the 1940 air raids and the texture of the ground was a mixture of sand, peat and gravel. The consulting engineers, under Sir Alexander Gibb, decided that it would be possible to build the units to their full height, but this involved further excavation and draining off the soil which had to be removed. Wells and centrifugal pumps were installed after a trench had been dug round the perimeter and the water was then pumped out and the soil removed. The sand provided an excellent foundation.

It was rolled out, covered with two thicknesses of ½in plaster-board and two layers of building paper. Concreting then began. Concrete was pumped direct to the shutters with 6in concrete filling pumps. A coffer dam was built while concreting proceeded and the basin was flooded on completion and the unit towed out into the dock.

On this site work proceeded rapidly. Mowlem's, Melville Dundas & Whitson took over on 28 October and excavation began on 2 November From 6–25 November well boring was in progress and the first concrete was placed on No 1 Unit on 5 January. Concreting was completed by 8 March and on 17 March the units were launched. The sheet piles were then re-driven, the basin pumped out and the site was ready for another batch of Phoenixes.

Towing out the semi-completed caissons also involved problems. Once the unit was floated off the casting bed, it could not be allowed to ground again or it would have broken its back. At first, when two units were built in one basin one unit was floated and towed out on one tide. Water then had to be drained out of the other unit with the next falling tide. The flood valves were then closed and the remaining water pumped out. On the next high tide the unit was floated out. But after experience in handling the first few units it became possible to tow out two units on the same tide. Usually tugs of shallow draught took the unit out of the basin allowing a more powerful tug to take it downstream.

Each construction firm built their caissons according to their chosen method. On some sites the casting bed was covered with hollow tiles and a layer of paper. The concrete was placed either by pumping, hoists, or with the aid of cranes. Large batching plants were provided on some sites; on others, smaller mixers were used on account of the limitations of the site.

Various forms of shuttering were used. They included simple wrot boarding with battened frames bolted through the wall; plywood was used instead of boarding, as were steel shutters, steel forms and various kinds of patent clamps.

Some idea of how the Phoenix building sites were distributed between the contractors and the maximum labour force involved on each site may be gained from the table overleaf.[22]

TABLE 2

Phoenix Construction Programme

Type of Caisson	No placed	Site	Contractor	Consulting Engineer	Total Labour Force (approx.)
A1	9	Tilbury Graving Dock	Balfour Beatty	Rendell, Palmer & Tritton	700
A1	9	Southampton Graving Dock, No 5	Pauling	Joint Engineers	700
A1	10	East India Dock (Wet)	McAlpine	Sir Cyril Kirkpatrick Group	1,400
A1	6	Portsmouth C Dock	Bovis	Joint Engineers	460
A1	6	Portsmouth C Dock	Laing	Joint Engineers	400
A1	8	Portsmouth Floating Dock	Lind	Joint Engineers	700
A1	4	Erith Basin	Nuttall	Oscar Faber	500
A1	4	Erith Basin	Costain	Oscar Faber	500
A1	4	Port of London Authority Basin	Demolition & Construction Co.	Oscar Faber	500
	60				
A2	4	Barking Creek Basin	Cochrane	Sir Alexander Gibb	500
A2	4	Barking Basin	Taylor Woodrow	Sir Alexander Gibb	500
A2	3	Bromborough Basin	Parkinson	Sir Alexander Gibb	450
	11				
B1	8	Surrey South Dock (Wet)	Mowlem	Sir Cyril Kirkpatrick	800
B1	4	Russia Dock Basin	Mowlem	Sir Alexander Gibb	490
B1	6	Russia Dock Basin	Melville, Dundas & Whitson	Sir Alexander Gibb	600
B1	4	Grays Basin	Gee, Walker & Slater	Sir Alexander Gibb	400
B1	3	Barking East Basin	A. Monk & Co.	Sir Alexander Gibb	400
	25				
B2	8	Stokes Bay Slipway	Holloway Bros.	Joint Engineers	800

Type of Caisson	No placed	Site	Contractor	Consulting Engineer	Total Labour Force (approx.)
B2	6	Stokes Bay Slipway	French	Joint Engineers	600
B2	6	Stone Point	Lovatt	Joint Engineers	700
B2	4	Langston Harbour Slipway	Trevor	Joint Engineers	600
	24				
C1	6	Goole Graving Dock	Boot	Joint Engineers	500
C1	9	Tilbury Graving Dock	Holland, Hannen & Cubitt	Rendell, Palmer & Tritton	700
C1	2	Cold Harbour Point Basin	Trussed Concrete Steel Co.	Sir Alexander Gibb	300
	17				
D1	9	Middlesborough Graving Dock	Harrison Group	Joint Engineers	500
D1	1	Woolwich Graving Dock	Parker	Oscar Faber	300
	10				

TOTAL 147

The time taken to complete a unit varied considerably. The A1 units, for example, took from two to seven months.

An idea of the magnitude of the task may be gained by quoting some of the quantities involved.

700,000 cu yd excavation
545,000 cu yd reinforced concrete
30,000 tons reinforcing bars of steel
6,000 standards (20,000 tons) of timber for shutters
15,000,000 lin ft of tubular steel scaffolding
50 miles steel wire rope
25,000 tons rubble for dock bottoms
80,000 sq yd hollow tiles for surface of sites
600 steel bollards
5,000 fair leads
150 sets towing gear
3,500 sluice valves and fittings.

Special provision had to be made to carry these loads and priority was given for Phoenix materials on the railways. Further problems included the provision of lighting to enable work to go on at night and shelter from air attack, while the accommodation of the workmen and their welfare demanded special attention.

As the programme advanced, the demand for skilled workmen, especially bar-benders and steel fixers, became even more acute. The Transport & General Workers Union therefore agreed that young trainees should work alongside skilled men. The latter, while doing their own job, had to supervise the trainee at the same time. The steel fixing force was increased in this manner by 900 men.

The numerous changes in design had, of course, delayed the start of the Phoenix programme but by the end of January concreting was going on apace. On 3 February the Prime Minister made an inspection of work in progress at East India Dock, but his enthusiasm for the advance that had been made in spite of the recent mishaps to the dock walls was not shared by some of the experienced foremen. Underlying their ribald comments on the heterogeneous task force was a shrewd appreciation of where the units would ultimately end and their wish was that most of the labourers should accompany them there.

Shortages of towing gear, valves and other essential equipment further delayed the completion of the Phoenix programme, especially in the Thames area. By the end of March, only eighty-three, or about 78 per cent of the total, had been completed and most of these were the smaller units. The Ministry of Supply anticipated that the programme would be completed by the end of April but the Prime Minister was not satisfied and at a special meeting held at 10 Downing Street on 25 April he ordered full pressure to be put behind the entire Mulberry programme[23]. In particular, there were to be no more modifications to the Phoenixes.

In fact, by early May, over 100 caissons had been completed and despatched to their parking places at Selsey and Dungeness. The remainder needed for the initial installation of the two harbours were ready by 23 May. The programme was almost two months in arrears, but considering the obstacles that had to be overcome, particularly the shortage of suitable labour, the contractors did well.

The task was not yet finished for the winterisation programme demanded more and stronger caissons. A force of 13,000 workmen was needed, though the requirement had reached its peak by July and afterwards no more than 7,000 men were being employed. By that time the V weapons were causing considerable damage to housing and the work force was switched to carry out urgent repair work. The whole Phoenix programme was completed by October 1944.[24]

The steelwork of the Bombardons was designed by the firm of Sir William Arrol & Co which was well-versed in landing craft construction. The organisation of the work differed from that of the Phoenixes in that one firm of steel erectors, Messrs Couzens & Sutcliffe, was responsible for drawing up the plans for the units in great detail. The work was then farmed out to twenty-eight steel makers who prefabricated about 2,000 steel tanks using 25,000 tons of steel of which, claimed the Admiralty, 80 per cent came from the USA. Some of the firms did not like being told what to do, but departure from the plans invariably brought them into difficulties, and they 'soon came to heel'.[25] All this work involved some 6,000 men while 2,000 more were employed on the assembly of the units at Portsmouth. This total included fifty welders and 320 steel erectors. The latter were nearly as difficult to find as welders because of the heavy demands on these men in the shipbuilding and heavy engineering industries. Another very important item was the moorings, which involved the Boom Defence Department in conjunction with DMWD.

Work on the Bombardons began at the end of October and by Christmas three prototypes were completed. By that time tests had proved the Lilos to be a complete failure.[26] From January to mid-May about twenty Bombardons were floated out from the dock each month and towed to the parking area in Weymouth Bay.

The shortages which have been described, such as the lack of skilled labour, the non-availability of the right kind of rolling stock to carry steel, and the numerous modifications in the design of the caissons, caused the whole Mulberry construction programme to be delayed by over a month. Meanwhile, the men who were to operate the equipment were impatiently awaiting its arrival. The Americans, who had no hand in its construction, inevitably found the delays more irksome than their allies. They were unaccustomed to the stringencies in men and material which a country at war for four years had grown accustomed to accept. At the same time they tended to come off second best when it came to the allocation of facilities and headquarters along the already congested south coast, then becoming the concentration area for Overlord. While they complained bitterly about the delays and about the complacency of British staff officers, with the exception of Hughes-Hallett, the British Mulberry contingent was demanding with equal insistence an opportunity to practise with their floating equipment. At last a date was fixed—for 6 June—and then the frustrated officers learned that the rehearsal had been postponed. They were, of course, unaware that the chance to put the equipment together would come under much more rigorous conditions than a mere exercise.

Fortunately for them, however, one trial of linking spans to a pierhead did take place at Selsey, largely at the insistence of Beckett who had been principally concerned in the development of the floating pier, and who was now attached to the British construction force as a 'technical adviser'[27]. He was concerned because Bruce White had come to the conclusion that it was unnecessary to carry a shuttle for each span in the manner previously described, and gave orders that the total was to be reduced to one shuttle per six spans so that the tows could be assembled more quickly. The demonstration was sufficiently convincing for Bruce White's order to be countermanded and the full complement of shuttles taken across the Channel. Unfortunately, Beckett was unable to convince the Americans to do likewise and, as will be seen, their failure to observe the proper drill for putting up the pier ended in disaster.

6 Towing, Parking and Mooring

It was the tugs which, in the end, proved to be the most critical item in short supply. Calculations for requirements were based on the time taken to cross the Channel; flood and ebb tides would influence the duration of the return trip; and time had to be allowed for bunkering, provisioning and watering before another voyage was undertaken. Experienced tug masters reckoned that about forty-eight hours would have to be allowed for a tug to cross the Channel and return (steaming at 4 knots on the way out). Unforeseen emergencies, inevitable in wartime, would probably increase that total by about twelve hours.[1]

But quite apart from the cross-Channel rôle, all the Mulberry units with the exception of the blockships had to be towed to their assembly places on the south coast. The delays in production made things worse because the towing commitments had to be compressed into a shorter period.

The first estimates were that 130 tugs would be required for Overlord, of which ninety were allocated to Mulberry. The British were to provide sixty-five and the Americans twenty-five.[2] But this figure hardly took into account all the tasks in home waters for which tugs were needed, far less cross-Channel work. To give some idea of how many British tugs were available, the total strength of tugs of 450hp and over amounted to 420, but nearly all these were fully occupied in duties other than those related to Overlord. True that there were greater resources in the USA but here there were conflicting demands from the Pacific.

When the Deputy Commander of Force Mulberry, Cmdr A. B. Stanford visited Washington at the end of 1943, despite a request signed by Eisenhower in his hand he had to 'steal, cadge, or persuade people to let go of tugs'. He felt it was a triumph when the number was increased to 200, of which 164 would be earmarked for Mulberry.[3] The US Army provided six ocean-going tugs of 1,225hp; fifty-four small tugs (ST class ranging from 400–700hp which would be suitable for placing the caissons and blockships) and fifty-five tugs of 300–400hp, while 120 small harbour tugs came from other sources.

The British meanwhile had set up a special committee to supervise the search for tugs, the result of which was that it was reluctantly decided to withdraw the rescue tugs reserved for the North Atlantic convoys. Also available to the British under Lease-Lend during the early part of 1944 were twenty-five tugs, including ten from the War Shipping Administration.

But many of the American tugs supplied were not up to the job.[4] They were inadequately equipped and their crews had had little nautical, not to speak of towing, experience. Some of them even lost their way en route to the beachhead and were captured by the enemy.

On the eve of D-Day, according to Tennant, only 132 tugs were available for service.[5] Of these twenty-four were required for towing different types of barge, leaving only 108 for towing Phoenixes, Bombardons and Whale. It was therefore necessary to set back the target date for the completion of the harbours on the far shore from D-Day plus 14 to D-Day plus 21.

The problem of tug availability was finally solved by the ingenuity of Stanford. He obtained a cameraman, a step ladder and several sets of children's bricks. At the headquarters at Portsmouth he 'chalked off the floor in four-mile swatches the full eighty miles to the far shore. Each hour I would move the bricks and snap a picture. In the end by simple counting I could tell the number of tugs in the Channel at any hour of any day. Eisenhower and the Prime Minister were delighted with the information and the unarguable scheme.'[6]

As the Americans were making such an important contribution to the tug pool, it was decided to put an experienced American tug operator, Capt Edmund J. Moran, in charge of the Allied tug control headquarters. This was set up at a signal tower at Lee-on-Solent. (Owing to the overloaded telephone exchange the staff had often to resort to giving instructions to the tugs lying out in the Solent by loud-hailer.) There was also a depôt ship for provisioning and repair. Moran was the right person to command the mixed force of tugs, comprising some ancient coal-burners and others driven by powerful diesels. In one day's visit to Portsmouth, Stanford writes, he was able to point out defects in gear which even at that late stage made all the difference.

One of the many jobs for the tugs was to tow the completed caissons to parking areas on the south coast. It was the original intention to moor them in suitable places but as the supply of moorings had been swallowed up by the need for Bombardon tackle, there was no alternative but to sink the caissons until such time as they were required when the water would have to be pumped out of them.

This unexpected development caught Tn5 by surprise. There was barely time to make alterations to the design, but valves were inserted into some of the cross-walls, and in the caissons still being built the compartments were reduced from ten to four. Holes to fit suction pipes were bored through the platform on the caisson's side and fitted with removable covers.

As already explained, the army had been made responsible for sinking the caissons. The sinking teams were given training first on the scale model in Northumberland Avenue (where Tn5 was situated); then they climbed over the caissons being built in the East India Dock to get the feel of their 'ship'. Finally, they were given instruction by members of the National Physical Laboratory at Teddington, not only in placing the units in position in the harbour, but also in parking them. Tables providing times in which the different types of caisson would sink were prepared by Iorys Hughes.

Gwyther and a colleague meanwhile investigated suitable parking sites.[7] The nature of the sea bed on which the units would rest was important as it was vital that they would neither sink into the mud nor break their backs on an uneven surface. Ideally, the sites should be near harbours where tugs could be berthed and repairs made. In the event, one of the sites chosen, off Selsey Bill, nearly proved disastrous because of the muddy sea bed. The second, at Littlestone-on-Sea, near Dungeness, was to be the reserve area but was later found to have bad scour in high seas.

The intention of Tn5 was that, provided the caissons were sunk in the right depth (which, in fact, they were not) they could be refloated by using valves for partial de-watering, tides for lifting, and pumps for floatation and trimming.[8] Four floats were assigned for the latter task, each carrying two pumps with 9in suctions, and each pump being capable of discharging 750 tons per hour. There were also six small floats, each fitted with four smaller 9in suction pumps able to discharge 130 tons per hour. As the large pumps could not empty the caissons because of the great suction head, 2in centrifugal bilge pumps would have to remove the residue from the bottom of the caisson at a speed of 100/120 gallons per minute. These small pumps would have to be lowered into the depths of the caisson and rested on cribs.

Thus the entire operation was complicated and time-consuming. Nevertheless because Tn5's original proposal of mooring had to be over-ruled it considered, wrote Rolfe many years later, 'that our duty was done when the units were taken from the building sites'.[9] As the months went by, it became increasingly obvious that in spite of assertions by Tn5 to the contrary, the pumping arrangements would be inadequate. Great store had been set by McMullen on two electro-turbine sets which were part of an emergency water supply system for London.[10] They were to be carried by Dutch schuyts (self-propelled barges brought over from the Netherlands by their crews) but as the Luftwaffe had recently launched a series of attacks against the capital these would probably be unavailable. Eventually, the Admiralty, on request, found five suitable vessels with pumping equipment to supplement the army craft.

The crisis broke in mid-April when the assembly areas were at last filling up with Phoenixes. One of the caissons being towed to Selsey broke loose from its tug and went aground near Littlehampton.[11] The tug crew flooded it to prevent it from drifting and endangering shipping but the spring tides did not allow refloating for ten days.

Capt J. B. Polland, RNVR, Deputy-Director of the Admiralty Salvage Service, was called upon to examine the caisson, though at that time he was unaware of its purpose. He quickly appreciated that it would be difficult to refloat in a hurry. The haste in which the units had been built had inevitably led to a number of defects. The apertures for pumps in the platform were sometimes blocked by steel bars. The wires operating the penstocks were either insecurely fixed or wrongly attached, and many of the apertures in the cross-walls were choked with debris, so preventing the water from escaping.

When Bruce White was informed, he sent Major P. B. Steer, a member of the caisson design team and before the war an engineer in the Port of London Authority, to make a report. Steer was accompanied by Polland, who by this time was privy to the secret. At Selsey they found great activity. DUKWs were plying backward and forward from the parked caissons carrying rations and ammunition to the gun crews living aboard them. Only the caissons, protruding from the sea like modern blocks of flats, were static. Polland quickly appreciated that many of them were in a similar state to the unit at Littlehampton. He also concluded that the caissons in deeper water would have to take priority over the smaller ones nearer the beach as they would take longer to raise. Moreover, the weather would be a critical factor because the salvage vessels would have difficulty in operating alongside the caissons. The two officers returned to London appalled at the task that would have to be tackled if eight caissons a day were to be despatched across the Channel from D-Day onwards.

The Americans, critical of the sang-froid, or optimism, of the British, which they contended was a form of burying their heads in the sand, were equally horrified. Fortunately, the US Navy was able to make a valuable contribution. One of their most expert salvage officers, Cmdre (later Rear-Admiral) Edward Ellsberg,[12] had recently arrived in London to take part in Overlord, having previously been engaged in clearing ports in the Red Sea and the Mediterranean of wreckage. Clark asked Ellsberg to see what could be done at Selsey. After inspecting the equipment on one of the War Office schuyts Ellsberg inwardly quailed. The centrifugal pumps being placed deep down in the hold were unable to suck up the water from the caisson over the side of the ship. 'Had the Royal Engineers,' wrote Ellsberg later, 'deliberately started to find the worst pumps possible for the tasks in hand, they could not have chosen better.' This was proved to be all too true when an attempt was made to raise a caisson.

Ellsberg at once reported to US naval headquarters in London, recommending that the job be taken out of the army's hands and given to a naval salvage officer, preferably American. Every spare salvage pump would, in the meantime, have to be sent to Selsey at once.

The British were slowly drawing similar conclusions. On 11 May Admiral Dewar, RN, Director of Salvage, told Mc-Mullen that so far none of the army's pumping equipment had been capable of shifting the parked Phoenixes and he doubted whether any of their equipment would be ready in time to be of any use. It was not until 21 May that the Chiefs of Staff, on the strength of a memorandum by Admiral Cunningham, decided to transfer responsibility for raising the caissons from the War Office to the Admiralty.[13] Cunningham estimated that only by using naval equipment, provided the army had done the preliminary work, would it be possible to raise three to four Phoenixes a day and so satisfy the minimum requirements of the assault force. If the army units reinforced the naval equipment, it would be possible to double that number, but the salvage crews would have to work day and night and they could not be expected to do that indefinitely.

A copy of Cunningham's memorandum was sent to the Prime Minister. After reading it, Churchill directed that daily reports on the number of Phoenixes raised should be sent to him and regarded the situation at Selsey with such gravity that he went down himself to inspect the scene of operations.

The conference on 21 May was followed by immediate action. Probably the most experienced salvage expert in the United Kingdom, T. MacKenzie, who had raised one of the largest German battleships at Scapa Flow after it had been scuttled by its crew following the German surrender in 1918, was appointed temporary Commodore. His task was to requisition every available salvage vessel in the country and get it to Selsey. Polland was appointed Chief Salvage Officer on the spot, Steer was to help him on technical matters relating to the caissons and in two to three days the situation was completely transformed. Men, craft and equipment began to pour into Littlehampton. As the rubber hoses quickly wore out through chafing against the concrete walls of the Phoenixes, Polland asked for steel pipes. These were cut into sections by gangs of Seabees and British sappers, both forces under Carline, who had come down from Garlieston to take charge of the repair organisation. They were then fitted into the apertures of the caissons and connected to the hoses.

By 27 May the Chiefs of Staff noted with relief that the pumping situation appeared to be in hand.[14] Ten naval vessels had now arrived (two were required to pump out each Phoenix) and five army vessels were either in position or en route to the scene. One of the British officers in charge of sinking the caissons, E. Haydock, contrary to Ellsberg's scathing comments, asserted that the work of the sappers with their 3in suctions assisted by fire-fighting pumps was exemplary.

Ellsberg, an old friend of MacKenzie, remained unobtrusively at Selsey to give advice if needed. Using the American salvage tug *Diver*, he discovered that some of the Phoenixes refused to become buoyant despite energetic pumping because of bottom suction.[15] In other words, they were stuck in the mud. The answer was to raise one end of the Phoenix at a time and blow jets of compressed air underneath to clear away the mud, thus allowing the water to seep in and destroy the suction. This operation had to be done with alacrity while the tide was coming in, increasing buoyancy but also threatening to swamp the pumps.

On 2 June, only four days before D-Day, Polland, exhausted but triumphant, reported that he would be able to meet the initial requirements for Overlord. Beyond D-Day plus three he could not foresee. There had been only three casualties. Two badly-leaking caissons had had to be beached and one had been sunk by the handling tugs. The experience had been chastening, but everyone had learned a lot which was to prove the utmost value on the far shore.

Difficulties had also been experienced with the Bombardons on account of the speed of construction. Unlike the Phoenixes, it was possible to conduct trials to discover how effective they were on breaking up the waves and to prove the all-important system of mooring.

Each breakwater was allotted a boom carrier, a net layer and nine boom defence vessels (their normal function was to lay anti-submarine nets across the entrances to ports). They now carried the marking buoys and huge anchors weighing from 3 to 5 tons. The first test took place off Newhaven.[16] The Bombardons were to be moored in two lines. The net layer marked the site, leaving the boom defence vessels to put down the anchors and secure them to the Bombardons. Pressure recording instruments were then lowered to the sea bed. The experiment revealed that the bolting of the units left much to be desired, confirming what Lochner had known all along to be the basic weakness of the design, but inevitable because of the lack of riveters and welders. Moreover, the couplings between two Bombardons sheared, causing a collision and swamping some of the tanks.

A further test took place in Weymouth Bay where tidal conditions and sea bed were approximately the same as those off the far shore. Lochner, when his design was subjected to much criticism on account of its alleged failure, took great pains to emphasise that the units were not intended to operate in depths greater than seven fathoms and that they could not be expected to be effective in winds over force 6. Arrangements were made that in the event of the Bombardons dragging their anchors and endangering other units explosive cutters, electrically operated from the shore, would free them. The test lasted for just over three weeks—the maximum period for which the floating breakwater was intended. It came up to expectation and withstood seas being blown by winds of up to 20mph. The instruments recorded a decrease in the height of the waves from 8ft to 3ft. Only two of the units became slightly waterlogged

After this experience the Bombardons were moored in pairs between buoys while the moorings were strengthened by the addition of clump anchors weighing from 5 to 8 tons. Instead of chain cables, the units were linked by thick manilla rope strops and were attached by steel couplings.

A final test was made at the beginning of May to find out whether these improvements increased the stability of the breakwaters. Although the sea remained calm throughout the exercise, the authorities were satisfied that adequate shelter would be provided in the event. Nevertheless, the commander of the laying force, Capt Charles Currey, RN, who had had a great deal of experience in setting up boom defences, seems to have been unhappy all along with his charges.[17] In the investigation made on the effectiveness of the Mulberries, he affirmed that the lee afforded by the breakwater was more spectacular than real and gave a false impression of strength. He never thought that it would withstand winds of force 5, far less force 6, or seas greater than 5–6ft high. Any attempt to strengthen the units would, he thought, be useless. Thus the breakwaters were only likely to be effective in fair weather.

Finally, a number of towing tests were made. In theory the units should have been no more difficult to tow than a ship but the prime purpose was to get them into position as rapidly as possible. They were therefore towed in pairs, in tandem, originally linked by wire hawsers, but when these gave rise to difficulties, the wire was replaced by manilla rope strops. These were much easier to handle and were used for the operation.

7 Building Mulberries and Sinking Gooseberries

On Sunday, 4 June, Capt Petrie, Brigadier Walter and other officers of Mulberry B construction force embarked in the cruiser HMS *Despatch* which was to be their main floating headquarters.[1] Also aboard was the Sea Transport Staff under Capt L. Thompson, an ex-Cunard commander, whose job it was to supervise the discharge of stores and personnel. Further down the Solent, now crammed with shipping of all kinds, were two wooden US submarine chasers which were to be the flagships of Clark and Stanford respectively. That evening a westerly wind was blowing, bringing rain with it. The landings had already been postponed for twenty-four hours because of the weather but early on the morning of the 5th Eisenhower gave the irrevocable order to go. The assault was to take place at high tide on the 6th, preceded, of course, by an immense minesweeping operation, followed by ships marking a safe channel with dan buoys.

No 1 Port Construction and Repair Group was responsible for building Mulberry B.[2] It had been formed in September 1942 by Lt-Col Stuart Gilbert to operate in North Africa. Gilbert, in addition to being an experienced port engineer, was a Territorial Army officer and had seen a good deal of active service. He was flown home from Italy in February 1944 and, to his great astonishment, was told to raise and train a special force. He demanded, and obtained, his old group, then in Naples. This provided a nucleus for the newly formed 969 and 970 Port Floating Equipment Companies handling the piers and pierheads (they were inextricably mixed together and operated as one) and 930 and 935 Port Construction and Repair Companies; the former was responsible for clearing the beach approaches and demolishing obstacles; the latter provided crews for the Phoenixes while being towed and parties to sink them when in position.

When the company commanders, several of whom, like Gilbert, had been serving overseas, first saw their men they were appalled to find that many of them, but by no means all, had been unloaded by units wanting to get rid of their trouble-makers and misfits. Gilbert relates how he arrived at Ryde one evening to find the commander of 969 Company, Major Ronald Cowan, formerly a civil engineer from Glasgow, taking a parade of nearly full company strength. Impressed by this keeness, he later asked Cowan how he had managed to instil such enthusiasm among his troops. 'They were all on defaulters' parade,' came the gloomy reply.

The task of turning this mixed force into soldiers fit for active service was not easy and there were only four months in which to do it. By D-Day the men had set fire to the officers' mess, ruined valuable equipment and there was a long list of pending courts-martial. A visiting colonel, recalled Cowan years later, watching an unhappy squad being trained to use the Bren gun remarked 'These men will *never* make soldiers; what a bunch of bloody goons.'[3] The name stuck, but in the event the 'goons' proved to be magnificent.

By the afternoon of 6 June, when the assault troops had established themselves on the far shore, the Mulberry forces were given the word to sail. First across the Channel, for Mulberry B, was group headquarters which had transferred to a Clyde paddle boat called the *Aristocrat* with a very shallow draught enabling her to pass over moored mines, while the teams of sappers allotted to build the piers and sink the caissons were carried in a variety of vessels, ranging from Thames and Channel paddle steamers to motor-boats and a boom defence vessel, HMS *Minster*, loaned for that purpose to the Americans. Parties were left behind to crew the pierheads, caissons and bridge tows when the time came.

Arromanches, in peacetime a quiet holiday resort, which was to be the exit for Mulberry B, was not one of the assault beaches. The engineer vanguard was therefore discharged farther east and had to make its way to the town on foot. Enemy opposition, apart from mines, was minimal; the coastal defences were manned by Ukrainians who put up no more than a token resistance. The first night was spent in an orchard outside Arromanches.

The Americans were less fortunate. Omaha beach, as it was

called, lay below three small villages, Colleville, St Laurent and Vierville, approached by sunken lanes which were to be the exits for Mulberry A. But on the morning of D-Day, unknown to Allied intelligence, the defences had been reinforced by an additional regiment.

The Americans had to fight hard to gain a foothold. There were also formidable beach obstacles including stakes and iron structures called Belgian gates, a number carrying Teller mines, the clearance of which set back the start of erecting Mulberry A.[4] The demolition work was carried out by the veteran 531 Engineer Shore Regiment of the US Army.

As far as the harbours were concerned, the first task was to make a hydrographic survey of the proposed site. Several hazardous reconaissances of the beaches had been made by parties of engineers landing from motor launch or midget submarine some weeks before D-Day and bringing back samples of beach and sea bed. But a more thorough investigation was required for the disposition of the floating and sunken units. In Mulberry B the navy surveyed up to the 5-fathom line at high water while the army took on the more dangerous area to high-water mark using echo-sound gear. Lt-Col Raymond (now Lord) Mais was in charge of the building of the piers. Early in his Territorial Army career he had transferred from the infantry to the Royal Engineers. He had seen service on the supply route from Persia into Russia as well as in other theatres of war, and though without conventional engineering experience was reckoned to be a 'rare pusher'; (he later became a Lord Mayor of London). By the early hours of 7 June Mais had established markers for the first two piers, the outer marker at high-water mark and a further marker for alignment purposes on the higher ground inshore.[5]

Out to sea the sites for the Phoenixes (placed farther out than originally intended to provide deeper water for deep draught vessels) and the blockships were marked with buoys. This was supervised by Lt-Cdr C. S. E. Lansdown, RN, from HMS Fernmoor, the soundings being taken from a motor launch. In Mulberry A this essential task was performed by Lt-Cdr Passmore, RN, aboard a small survey craft which he had navigated across the Channel.

On 7 June the blockships, which shortly before D-Day had assembled in Poole harbour, were now steered towards their final resting place. (See Fig 19.) Scuttling was not easy.[6] The ships had to be held in position by tugs broadside on to a fast-running tide with the wind usually in the wrong direction. The charges were then exploded allowing the water to come in and the ship to settle on the bottom. An important lesson learnt as the scuttling proceeded was the need for one ship to overlap the next one. When spaces were left the sea pounded through causing excessive scouring on the bottom which, in turn, made the ship settle and, sooner or later, disintegrate.

Gooseberry I was on the right (American) flank protecting the beach known as Utah.[7] The operation, directed by Stanford, was interrupted by fire from German 88mm batteries. *Wason*, the first ship to go down, swung round as she was settling. Apart from the difficulties mentioned above, the mishap was due to the skippers of the tugs (old coal-burners flying the Red Ensign) who cut loose too soon in their anxiety to steam out of danger as quickly as possible. The German guns sank the second and third ships approximately in the right position, though slightly too far apart; their destruction was later confirmed by Berlin radio.

Stanford turned this unforeseen setback to good account and decided to make two crescent-shaped harbours based on the ships out of line. The remaining ships were sunk accordingly. By 13 June Gooseberry 1 had been completed. It proved its worth as a respectable tonnage of supplies was unloaded within its shelter. Eventually it was capable of accommodating 75 Liberty ships and an armada of smaller craft. Two cause-ways, made out of NL pontoons, were laid, one of them extending beyond the Gooseberry. These enabled troops to land dryshod and small craft could unload alongside them.

Gooseberry 2 provided the first shelter for craft discharging in Mulberry A. In order to allow rapid access to the beach for light craft, the Americans decided that two gaps should be left between the ships, in effect making three breakwaters. Royal Navy officers had warned that the sea would pour through the gaps in bad weather and they were, unhappily, proved to be right when a storm blew up on 18 June. Cdr C. R. Dennen,

111

USNR, directed the scuttling operation and the great bulk of *Centurion* provided a cornerstone to the breakwater. The next three ships, *James Iredell*, *Baialoide* and *Galveston*, were sent to the bottom under sporadic gunfire from inland. However, the Gooseberry was completed by 10 June. Here, too, a pontoon causeway was erected by the Seabees opposite the St Laurent exit and was in use by that date.

The third Gooseberry at Mulberry B was planted by Lt-Cdr A. M. D. Lampen, RN. Working closely with him was Capt John Jellett, RNVR (Special Branch), recently flown home from Malta and appointed Superintending Civil Engineer for Mulberry B (he was, in effect, the naval counterpart to Gilbert). After the survey, Jellett decided to realign the line of blockships at a more northerly angle because at one point the water was deeper than anticipated.[8]

The *Alynbank*, the first ship to go down, lurched out of control, the tugs having failed to hold her against the strong current, thus prolonging the sinking operation for more than the stipulated four minutes. She ended up at right angles to her intended position and the gap was later filled by three caissons. As it happened, the error provided a bonus of 110 acres of sheltered water which was used by shallow-draught vessels. The remaining blockships were sunk much more skilfully with a good overlap between each vessel. This required a high order of seamanlike handling of tackle on the part of tug and block-ship crews.

Gooseberry 4 provided shelter to Courseulles, seven miles east of Arromanches and which had small but undamaged harbour installations operated by No 2 Port Repair and Construction Group. Offshore shoals made it possible to combine artificial shelter with nature—an ideal situation.

Finally Gooseberry 5 was located off Ouistreham on the left flank of the British sector. Of the nine blockships, six were merchantmen and the remaining three were the naval vessels *Courbet*, *Sumatra* and *Durban*. Three weeks after the landings an intercepted German radio message reported *Courbet* grounded, believed damaged; the enemy evidently had not appreciated the purpose of the Gooseberries though artillery fire was later directed on the beach area making the harbour untenable. But this loss was more than compensated for by the volume of supplies discharged through Courseulles.

In spite of the engineers' doubts, the blockships were an asset, particularly after the decision to enlarge the beachhead. Never before in naval history had they been used on such a scale. From a technical point of view the scuttling procedure was criticised as having taken too long and on the grounds that more powerful charges should have been used and placed lower down in the hold. The Admiralty's answer was that the holes in the ships' sides were deliberately kept high in order that the ballast should not be washed out of the hulls in bad weather.

The boomlaying craft to mark and lay down the moorings for the Bombardons set sail on the night of 6 June. Two events happened which the protagonists of these units believed lessened their value. Admiral Ramsay decided at the last moment to reduce the layout of the Bombardons at both harbours from a double line, as originally intended, to a single straight line. This decision had the effect of increasing the mooring stresses while at the same time reducing the amount of wave reduction obtained. Secondly, the units were moored in error in 11–13 fathoms, whereas the moorings had been designed for a maximum depth of 7 fathoms.

At Mulberry A the Bombardons were laid by Cdr L. D. Ard, USNR, an adept sailor whose experience stood him in good stead with these clumsy units. At Mulberry B the Bombardons were laid under the direction of Cdr C. I. Horton, RN. Both British and American floating breakwaters, each a mile long and consisting of twenty-four Bombardons, were completed by 17 June. The Admiralty claimed that in Mulberry B the Bombardons protected the harbour from the north-west reducing the wave height by about 50 per cent while the building was in its critical

XIII Bombardons form a floating breakwater at Mulberry A

phase.[9] They also claimed that the Bombardons in conjunction with the blockships enabled a great mass of men and equipment carried by DUKWs, Rhino ferries and barges to be landed in the face of almost consistently bad weather (see Appendix) during the first fortnight of the landings. The staff requirements, they considered, had therefore been met.

Towing of the Phoenixes across the Channel began on 7 June. Each unit carried a gun crew (usually four men) and two sailors, one of whom was to keep in touch with escorting vessels. They were able to shelter from the weather in a corrugated iron lean-to on the side platform.

At Mulberry A the first two Phoenixes were not in position until 10 June because of enemy fire against the harbour area. But four days later thirty-two out of fifty-one units were sited and already providing a useful breakwater.[10] As with the blockships, sinking was not always easy and several units broke away from their tugs.

At Mulberry B planting began on the 10th at what was called the detached mole to the west of the northern entrance. This

XIV Fixed (Phoenix) breakwater at Mulberry B

was swung further out to sea than planned providing about thirty-five acres for vessels with deeper draughts than coasters. The army had nonetheless to press for extra space. When the Prime Minister arrived at the end of June to tour the beach-head the cruiser in which he had crossed the Channel was berthed snugly inside the western entrance. Walter therefore suggested to Hickling, who had superseded Petrie as naval officer-in-charge of Mulberry B, that if he could berth Mr Churchill, he could berth an additional Liberty ship. The eastern arm was also modified because of changes in the blockship lay-out already mentioned.[11] Again, units were planted in deeper water than planned. By 18 June the detached mole was completed. A total of twenty-five caissons had been placed and only one was out of position, due to a tug colliding with her charge in the dark.

No more than four Phoenixes were lost in the assault phase, taking most of their crews with them. Two were sunk by enemy action (one by a torpedo and one by a mine) and two were lost in bad weather.

Installation of the stores and LST piers was perhaps even more complicated than scuttling blockships and sinking caissons. In Mulberry B the tows began to arrive on 9 June,[12] and from then on, until the piers were completed, Ronald Cowan's men worked day and night. At night they were hampered by the dense smoke screen protecting the area and during the day the 18in shells from HMS *Rodney* and USS *Utah* screamed like express trains overhead on their way to targets around Caen. At least they were not bothered during the day by enemy aircraft due to the almost overwhelming Allied air superiority.

Capt John Luck, RE, an ex-Thames tug skipper, directed the steam tugs towing the spans into position. Five motor towing launches (MTLs) belonging to 334 Harbour Craft Company of the US Army also took part under the charge of a warrant officer named John Heming and his sergeant, Max Hartdigan, from Michigan and San Francisco respectively. These boats were manned by a crew of three; they were more nimble and, because they burned oil, had a greater range than that of the steam tugs. When the Americans were at Cairnhead, they had seen how unmanoeuvreable the coal-burners were when it came to linking the spans and, unknown to the British, made provision for a flotilla of MTLs to be allotted to the British harbour in addition to their own.

Collaboration between the two forces was excellent, unlike the sometimes acrimonious exchanges during the preparatory period. Walter and Heming, for example, became great friends and the brigadier was often to be seen aboard the American launches arguing with the crews on the merits of baseball teams. Cowan relates:

> Though we never found out how, Walter had expert knowledge. One wet, choppy afternoon, with the brigadier aboard, Heming and Luck completed a particularly difficult coupling-up operation with a partly submerged tow. The sappers waved and cheered as the free bridge end slid into the socket (there were only inches of tolerance). These salutes were normally acknowledged by the Yankees but on this occasion we could see that an apparently bitter argument was taking place on the deck of the MTL between the brigadier and Heming. As Wally Walter leapt off the launch on to the pier he was shaking his head.
>
> Distressed at this apparent breakdown of Anglo-American relations, I said something about the expert way the Yankees had handled the tow. The brigadier looked at me sadly: 'Yes, yes, I agree', he said, 'but I am worried about that fellow Heming. He knows *nothing, nothing* at all about baseball', and off he went grinning down the pier.

The choppy seas made it difficult for the tugs to manoeuvre the bridging into line. Luck and Heming had to judge the exact moment to slot the free span into the trumpet guides at the bridge end. Miscalculation could maim, even kill, the bridge crews (one officer was knocked unconscious by a girder one rough afternoon). Such was the mutual trust and co-operation that by the morning of 14 June the (east) stores pier had been laid to a distance of about three-quarters of a mile from the shore and a spud pontoon attached to it. Discharge from coasters into 3-ton lorries began at once.

Especial care was taken to make the floating pier secure. This was due in no small measure to Lt-Col J. R. Sainsbury, RE. He had been associated with Whale from its early development at Garlieston and shortly before D-Day was attached to

Mais's headquarters to ensure that installation was carried out properly. Sainsbury had, in addition, been convinced by Beckett that the full complement of shuttles should be taken to Normandy. Sainsbury was supernumerary to Mais and, finding himself superfluous, set himself the task of ensuring that the floats were properly secured, as he observed that in the heat of the moment only alternate floats were being anchored. With a sapper section borrowed from Cowan he not only moored each float from both ends but secured each float to the other, while those nearest the shore were fixed by anchor bolts to the sea wall. Luckily there was ample wire to make this possible. These precautions were completed before the storm and Beckett, who was also helping to supervise operations, remarked to Sainsbury that the pier was 'braced up like a woman's corset!'

So that there should be no delay in discharging, priority was given to getting a circular flow of traffic going. Mais was therefore ordered to concentrate on completing the second, or centre, pier of the stores pier and this was done by 7 July. Gilbert decided after the storm to turn the third, or west, pier into a discharging jetty for barges as they were proving so useful. Among the casualties on the cross-Channel journey were some of the inshore floats equipped with legs to prevent them from being damaged by rocks on the shore; others failed to arrive for some unknown reason. Mais decided to use ordinary steel floats instead and restrict traffic over the period when they were drying out.

Installation of the LST pier was also delayed by the casualties to the floats. One reason for these losses was the failure of the hitherto reliable weather forecasters (Gilbert asked for one to be sent to Normandy). The result was that tows set sail and had to return to port, otherwise they would have been waterlogged. The concrete floats, which even in sheltered water had proved vulnerable to flooding, were unable to cope in mid-Channel; and being towed broadside on, the concrete bollards were subjected to great strain, the tugs often pulling them out by the roots. Sometimes the erection tanks were insecurely fixed and came adrift when the spans were towed above the regulation speed (as happened when E-boats were reported in the vicinity!). This caused the front span to sink. The fairings, designed to prevent the floats from being swamped, also occasionally came adrift, damaging the sides of the floats.

In the end, about 40 per cent of the tows were lost. By 13 June, after five Whale tows had been sunk, Tennant decided to shorten the tows to three spans and to sail them in daylight rather than at night.[13] Fortunately casualties were small but for the sappers who had to man the spans it was a wretched task. They had no more than a tarpaulin to shelter them and for most of the voyage they were cold, wet and sea-sick, to the accompaniment of the screeching, groaning and grinding noise of the sections as they worked in the seaway.

While the pier building was in progress 930 Port Construction and Repair Company under Major N. W. Smithson, was clearing the approaches to the shore ramps.[14] Beehive charges cleared exits for the piers and for the DUKWs leaving the beach. A neat hole was bored 6ft deep and 6–8in in diameter, into which a large charge of gelignite was placed. The sappers also cleared the area of mines and booby traps. As the shore ramps for the LST pier could not be brought far enough in-shore 935 Port Construction and Repair Company under Major A. H. Hinrichs put up Bailey bridging to surmount the gap.[15] Traffic to and from the stores pier entailed the building of a flyover bridge and this required the demolition of a number of buildings near the front. The mayor of Arromanches agreed, without reservation, that the plan should be carried out and most of the work had been completed by the end of June.

The Americans experienced a more dangerous start to their pier construction as it was within range of snipers and there were more obstacles than in the British harbour area. Although tows began to arrive on 9 June erection did not start for another three days. Clark therefore decided to concentrate on completing the centre pier, combining a stores pier with an LST pier, as it was important that tanks should be brought ashore to cut off the Cherbourg peninsula and advance inland. Work now went ahead with the utmost speed under the direction of Lt W. L. Freeburn and his assistant, Chief Bos'n F. F. Hall. According to Stanford they had skilled mechanics under them whose work was 'super'.[16]

While there was no doubt that the work of the Seabees was good, their initial training on the equipment in Scotland had begun too late for them to benefit from actually linking the spans and anchoring the floats themselves, as this task had been completed by the British. Moreover, the officers who had been sent on the course did not seem to appreciate, in the way that the British did, the need to secure the floats properly. Later, they turned a deaf ear to Beckett's warnings, and training before D-Day was, in any case, limited to coupling a few spans.[17]

For this reason, on the far shore the Americans thought that the mooring of alternate spans would suffice, with the result that when the bad weather came the anchors dragged and the bridge swung out of position. The British, on the contrary, doubled their moorings when the gale warning was given.

It may well be true that the American harbour was more exposed than the British and the sea bed may have been less suitable for anchoring. But there can be little doubt that the Americans had less respect for the weather than the British, reinforced by their greater familiarity in operating the equipment in all weathers.

Like the British, the Americans suffered from the loss of tows in the Channel. Ellsberg was already on the beachhead waiting for orders to clear the port of Cherbourg. Clark asked him whether it would be possible to use the lighter floats, intended to carry a 25-ton load, for the Sherman tanks of the 2nd US Armoured Division weighing 38 tons. Or would they submerge under the strain? Ellsberg, who had a slide rule with him, set to work to measure a float and then spent most of the night alone in a cabin in Clark's headquarters ship working out the problem.[18] He was able to reassure Clark. 'The 25-ton pontoons would remain afloat, though only by an eyelash, under a 38-ton load, but I would guarantee they would remain afloat.' It was, of course, essential that two tanks were never simultaneously over the same span. Ellsberg himself, walking backwards over the half-mile to the shore, supervised the passage of the leading tank on 16 June when the centre pier was completed and joined to the pierhead.

A second pier and pierhead were ready two days later.

Eleven LSTs docked and discharged taking 64 min per ship as opposed to the 12–14 hr spent waiting for the next high tide. The average time of discharge, states Stanford, was 1.6 min per vehicle.[19] Artillery which would have been difficult to haul over the soft sand was also landed.

It must not be overlooked that until the piers and pierheads were completed stores and equipment were being brought ashore by the DUKWs, Rhino ferries and barges. There were about 115 Army craft, barges, lighters, ferries and 100 DUKWs being used in each harbour. The DUKWs especially became virtually indispensable to the success of the operation, operating from D-Day onwards despite the rough weather. They not only moved 'cargo from ship (often three miles out to sea) to shore, as other ferry craft could, but (transported) that cargo overland to a dump.' They went far to remedy the shortage of cranes and trucks in the early stages of the assault and 'what might have been a random piling of supplies on the beach (was converted) to an orderly movement from ship to dump.'[20] The loads, mainly ammunition and petrol, were carried in three nets, the contents of which were transferred into lorries on shore. The DUKWs were not, of course, able to handle some of the larger, awkward loads, like Bailey bridging sections, included in the cargoes.

By 16 June the enemy must have appreciated the significance of the Mulberries for he began to launch air attacks in the area.[21] In addition to the great Allied air superiority over the beachhead and beyond, the anti-aircraft defences at sea and ashore were on a prodigious scale and more often than not were unable to discriminate between friend and foe. The guns were supplemented by balloons and smoke screens laid by trawlers at dusk and dawn. As the shipping was fairly widely dispersed and there were no warehouses to bomb, the Germans put all their effort into minelaying. They concentrated on the harbour entrances, but mines laid anywhere off the beaches were bound to be set off sooner or later, owing to the intensity of the traffic.

In the early hours of 19 June enemy aircraft began to drop pressure or 'oyster' mines which were detonated by the pressure wave under a ship's hull moving through shallow water. Two of these mines were discovered intact on dry ground and shipped to Portsmouth for dissection. Their use had long been anticipated and although there was no fully effective counter-measure, by reducing a ship's speed it was possible to give some measure of protection. Most of the mines were set off by the sea swell.

Mine-spotting parties were now placed round the harbours. Their job was to mark the spot where a mine entered the water so that it could then be detonated by an underwater explosion. A prize of a bottle of Scotch was offered for locating a mine. One evening a mine was reported to have been dropped in Mulberry B but Hickling, unwilling to stop work, announced that it was a dud shell. Next morning he heard an explosion and, to his horror, saw the plume of a mine with a DUKW sliding down it. Fortunately the DUKW survived and was driven ashore, whereupon Hickling, rising to the occasion, signalled its commander his surprise that DUKWs now included minesweeping among their varied accomplishments.[22] If the crew reported to his headquarters they would receive the prize.

Later two rather sheepish RASC drivers appeared and the driver told Hickling that they had heard about the mine while going out to a Liberty ship. 'I told my chum,' he continued, 'we'd better keep clear of that there mine, when all of a sudden there was a nasty bang at the back.' When the pair received their reward the driver asked Hickling, 'Sir, can you tell us where there is another mine?'

During the first two weeks of operations about 27,500 tons of supplies had been discharged through the two Mulberries, though no vehicles had been landed at Mulberry B as the LST pier was still incomplete.[23] The planned target of 6,000 tons per day which should have been reached by 20 June was not achieved until 6 July. The maximum number of vehicles landed in one day at Mulberry B was 1,225 but this did not occur until 18 September.

At the same time the discharge of stores by coasters and landing craft from the beaches protected by the Gooseberries, particularly at Courseulles, was much greater than had been expected and no damage was done to their hulls; the only drawback was the need to wait for the next high tide before being refloated. Detractors of the Mulberries used the following statistics as an argument against their necessity.

TABLE 3

Discharge of Supplies, Vehicles and Troops, 6–19 June 1944

	Supplies (long tons)	Vehicles	Troops
British Beaches (excluding Mulberry B)	120,729	50,400	286,586
Omaha	74,563	29,242	205,762
Utah	49,841	14,344	126,507

The British and American beachheads were now linked up. Twenty divisions were ashore and had a numerical superiority over the enemy. But the build-up of supplies, particularly ammunition, continued to be vital and from mid-day on 19 June this suddenly became a matter for doubt.

8 The Storm

On 16 and 17 June the Channel had been too rough for towing operations. On the evening of the 17th an overdue spell of fine weather appeared likely and Admiral Tennant ordered a maximum effort for the evening of the 18th.[1] Four Phoenixes and twenty-four Whale tows were despatched to the beach-head, but before they sailed a cold front from Iceland was already spreading rapidly over the British Isles. At the same time a depression from the Mediterranean moved northwards over France. These two disturbances combined to produce a north-north-west wind, later veering to north-east, of force 6–7 during 19–20 June and moderating only slightly on 21–22 June. The wind swept along the exposed north French coast reaching gale force and the sky was overcast with thick low cloud.[2] Eleven of the tows were sunk and one of the Phoenixes ran aground near Omaha beach.

The more experienced sailors in the Mulberries, looking at the sky, tended to be sceptical about the weather forecast. At Arromanches, Walter, after a discussion with Petrie, instructed his force to take immediate precautions.[3] As the Whale equipment with its moorings and thin-skinned floats was the most likely to suffer damage, Mais issued the following instructions to his officers. First, all moorings were to be doubled; secondly, all shipping on the weather side of the piers was to be moved so that they did not drag anchor and bear down on the piers and pierheads. Thirdly, the pierheads were to be spudded up to their fullest extent. Fourthly, the tugs were to be stocked with at least two days' rations and fuel and they were to be continuously manned. Some of them were to remain to windward. If a vessel or Bombardon were to break loose, they were to get a line aboard it and tow it clear; otherwise they were to sink it with PIAT (anti-tank projector) fire. All ships in the harbour, with the exception of those actually discharging, were warned to up-anchor at once—or else! A few bursts from Witcomb's Oerlikon gunners on the pierhead drove home the message.[4]

Early on 19 June the wind increased in violence. The troops were already tired when it started. They were beyond ex-

123

haustion when it finished. Priority was given to saving the stores pier; the east roadway being the nearest to completion was the most vulnerable.[5] By the next day waves at least 8ft high were crashing against the pierheads and, as Cowan relates, 'washed over the floating bridges, making them buck and rear like wild horses, so that the men had to cling to the steelwork or be washed away to drown—and this went on, hour after hour, day and night. In the wild daylight and in the dark, out of the gloom would appear crewless, drifting vessels— pursued always by the British and American tug crews. In the darkness men had to jump from the tugs and from the piers on to these abandoned craft to attach tow lines so that they could be hauled away. There were some incredible (and unsung) feats of seamanship and daring by Luck and Heming and their men. The Schermuly rocket pistols, for firing light lines, were absolutely invaluable.'

For the stores pier, the crisis came before nightfall on the 21st. A steel float, according to Cowan, had become wedged

XV Effects of the summer storm. Damaged pier at Mulberry B (*Public Record Office*)

XVI Effects of the summer storm. Span of LST pier washed ashore. It is being refloated using erection tanks (*Public Record Office*)

underneath one of the spans. With each wave it crashed against the steel trusses and was smashing the end of a concrete float. Four men, including Lt W. Rigbye, recently commissioned, had somehow managed to jump on the wet and greasy steel deck of the float. Rigbye hurt his ankle and was unable to move; he clung on with the waves breaking over him. As the wind and sea drove the float still further beneath the bridge girders, it was only a matter of minutes before Rigbye would be crushed. In the half light Mais and another officer leapt from the bridge on to the float; they passed a line to one of the tugs and the float was towed away. Rigbye was rescued and the bridge saved.

The LST pier, only half built, did not fare so well and caught the brunt of the gale. Girders were twisted and several spans, including a telescopic span, parted company with the bridge. One was washed up against a cliff. Sappers bulldozed a path to it, jacked it up and floated it off with the help of an erection tank. Some of the rag bolts connecting the steel transoms to the concrete decks of the floats were torn out and the latter became waterlogged. Exposed reinforcing bars punctured the rubber buoyancy bags. Later, the floats were filled instead with empty oil drums, cans, and even ping-pong balls.

The two pierheads were practically isolated. Witcomb and his men had spudded up their pontoons to maximum so that

XVII Effects of the summer storm. A severely-damaged Phoenix

the warning lights on the control panel were permanently red. In spite of this the legs failed and dragged on the sea bed, a situation in which the spud ropes could break or the spuds be torn from their keepers. But the pierheads held.

The wind drove the sea over and into the tops of the Phoenixes filling some of them up to the brim. The water could not escape, not only because the wooden hatches over the apertures on the side of the caisson had not been removed, but the penstocks inside at the base of the unit were designed to let water *in* at a controlled rate rather than out. When the tide went out the internal pressure of a 30ft head of water caused five units to split. Another was shifted from position by scour. All the damaged units were in the detached mole. The block-ships, save for two which broke their backs, held fast; they were probably saved by being placed on a sea bed of rock.

Away out to sea and more exposed than anything else, the line of Bombardons disintegrated altogether. Some of them broke loose and drifted shorewards and had it not been for the tugs on the alert would surely have crashed against the piers. The Admiralty report concluded that their failure was due less to faulty moorings than to the working loose of the bolted connections.

Apart from the Bombardons, about 250 small craft, including several Rhino ferries, were sunk or battered on the shore. The MTLs all survived because, in the first place, they kept on the move and, secondly, their crews, conscious of operating under the British, were on their mettle. The American MTLs were all lost. Despite the storm, vessels continued to unload and white blobs were painted on the pier girders to guide the trucks in darkness. On 22 June, for example, 1,200 tons of ammunition out of 1,500 tons for the British forces were unloaded over the stores pier, usually prohibited for this purpose because of the potential danger from air attack. Above all, thanks to the 'bloody goons' the harbour was saved. It was their morale, wrote Walter afterwards, that 'built Mulberry and not merely technical skill.'[6]

At Mulberry A no less determination was displayed by the sailors and Seabees, but being so exposed it received the full brunt of the storm. By ill fortune, as well as being inadequately moored the piers were struck by drifting LCTs and other craft. Like the British, Clark concentrated on trying to save the piers and pierheads.[7] Already exhausted by urging on his men while under fire from shore, he wore himself out and was compelled to return to England, his place being taken by Ard, now bereft of his Bombardons. One 'of the piers was completely ruined. Its centre span of bridging was bent and twisted in a

XVIII Effects of the summer storm. Drifting Bombardon (*Imperial War Museum*)

great arc curving to the west, and its Beetles were either broken loose or beached, or smashed and filled with water. The other pier was not as badly damaged, but was also bent in a great arc.'[8] On the night of the 21st the spuds of two pierheads broke and the pontoons were carried ashore. Stanford now concentrated on saving the remaining pierhead. The clutch was released allowing the pontoon to ride freely up and down the spud legs. In this way, it survived, rising and falling with the swell, its decks cluttered with sleeping men alongside the dead who had been pulled out of the water.

But it was the destruction of the Phoenix breakwater that put an end to any chance of saving the harbour. Out of thirty-one units twenty were destroyed; three of them having their backs broken through scour; the others through having their sides burst open by the water pressure within. Several were hit by Bombardons descending on them 'like a damned great catherine wheel' according to one witness. Seven blockships broke their backs and others settled deep into the sea bed.

Well over 100 craft were lost and discharge of supplies was brought to a standstill. Vital ammunition supplies had to be flown in and several ammunition ships were beached for unloading. In spite of all this, as soon as the storm abated the DUKWs and ferries began to operate again. On 23 June 10,000 tons were unloaded at Omaha and shortly afterwards normal output was being achieved, though without the advantage of the piers.

The Gooseberries protecting the other beaches survived. Utah, also being exposed to the west, lost two blockships and three others were in a poor way. Shipping tended to be diverted to Grandcamp and Isigny. To the west, at Ouistreham the three warships sank deeper into their sandy bed; *Sumatra* had received a torpedo amidships; *Courbet* was weakening and one of the merchantmen had a broken back.

The storm now spent, the commanders of the respective harbours inspected the damage. Mulberry A had undoubtedly suffered the greatest.[9] Yet such was the spirit of the construction force that it believed that, despite the chaos, the port could be restored to working order. Ellsberg, no inexperienced salvage officer, supported this view. But the Mulberries had always had their detractors. Commodore W. A. Sullivan, Supervisor of Salvage, US Navy Department was one of these. Moreover, he and Ellsberg had not, on a previous occasion, seen eye to eye. After a cursory inspection Sullivan reported to the senior naval commander, Admiral Kirk, that a salvage operation was impracticable. Kirk accepted his verdict, influenced by some of his staff who were scornful about the force of the gale, implying that the harbour itself was inadequate.

Eisenhower therefore ruled that Mulberry A should be abandoned but that the Gooseberry should be reinforced with additional, stronger caissons. Mulberry B, meanwhile, was to be completed to the original plan and strengthened to last at least until October, if not to the end of the year. The fateful Bombardons were not to be replaced. Gaps were to be filled by salvaged equipment from Mulberry A.

Eisenhower's orders had already been anticipated. After the storm Mais had driven to St Laurent in his jeep and spoken to Col Harry E. Bronson, an old friend of his, in command of the beach engineers.[10] Although disconsolate at being unable to complete their harbour, the Americans began at once to collect spares and salvage undamaged equipment for towing round to Arromanches. Bronson also lent the British two sets of cutting and welding equipment.

It was now possible to repair the LST pier and young Rigbye, supported by crutches made from two broomsticks, was put in charge. With the assistance of a 60-ton floating crane which had arrived at Arromanches, bulldozers, and all the available welders and steel erectors, Rigbye had the LST pier in operation by 17 July and it immediately began to discharge tanks and vehicles.[11]

Reserve supplies of floats had been coming across, meanwhile. Experiments were made in carrying them in a Landing Ship Dock (*Northway*) and one long section of roadway was shipped over in two sections on a car ferry. The other piers were now protected by kapok foot bridges against drifting craft and ignorant boat crews who wanted to tie up to one of the floats. At the breakwater divers removed the wooden hatches on the Phoenixes and sappers cut holes into their sides with pneumatic drills to let out the water.

Eventually, the breakwater was reinforced by new and much stronger units which double-banked the original line. Some of them were built with square ends, as speed was no longer essential for crossing the Channel, to prevent scour. The tops were roofed over with steel sheet piling. To give them more stability the open units were filled in with sand by two dredgers *Leviathan* (10,000 tons) and *Burbo* (6,000 tons). In order not to disturb the neighbouring beaches they had to bring the sand from about seventy miles up the Cherbourg peninsula. Rubble and stones were poured into gaps and held together with wire netting to reduce the effect of scour. The *Vestra*, a port repair ship and, in effect, a veritable floating workshop, was a great stand-by during this phase of the operation.

9 The Task Completed

The harbour at Arromanches, now no longer called Mulberry B, did not lose its importance until late in the autumn of 1944. The breakout from the beachhead and the advance across the Seine to the German frontier imposed a severe strain on the supply system. Food, ammunition and petrol were the three main prerequisites. But Cherbourg and the Brittany ports, as well as those east of the Seine, took much longer to get into operation than anticipated because of enemy demolitions and minelaying. And even though the great port of Antwerp was captured intact, it took weeks to clear the approaches.

Thus at a vital period of the campaign the supply lines, particularly those of the Americans, were dependent on Cherbourg, the Arromanches Mulberry and the Normandy beaches. A one-way traffic system known as the Red Ball Highway was set up which was debarred to all civilian and local military traffic. Along it convoys 'swept along at high speed day and night, in an unending stream.'[1]

Of special importance was the need to berth the Liberty ships able to carry over 7,000 tons, three times the total of the largest coasters. The latter were used less and less as winter approached. The Arromanches Mulberry was the only sheltered anchorage between Cherbourg and Le Havre. By mid-October American, as well as British, Liberty ships were discharging at Arromanches. Although it had been intended to close down the Mulberry at the end of that month, it remained open until 19 November.[2]

Apart from the first half of August, the weather continued to be uncertain right into the autumn. Fog and high winds were experienced late in July and another severe storm interrupted discharge of men and supplies on 1 August. Heavy seas pounded the coastline at the beginning and end of September so that strengthening of the Mulberry and the Gooseberries at Omaha and Utah assumed great importance. The new Phoenixes being built around the English coastline were awaited with impatience but the V weapon offensive, which began soon after D-Day, drew away a number of workmen for emergency repair of

131

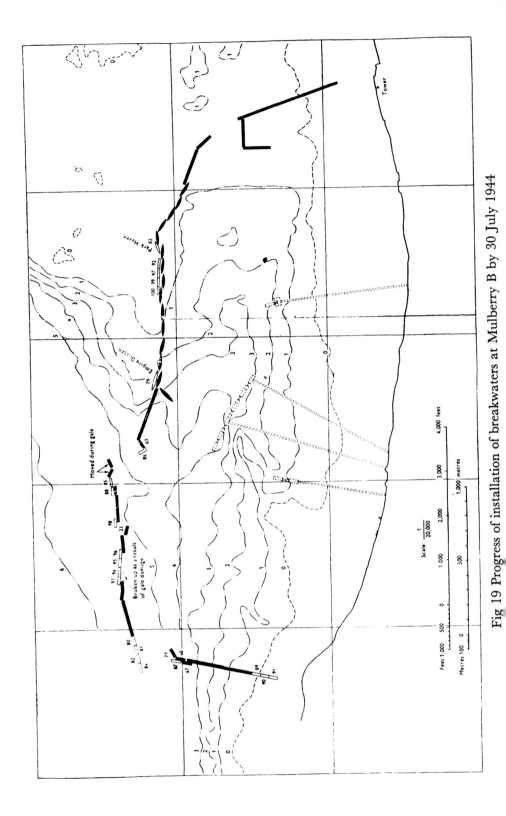

Fig 19 Progress of installation of breakwaters at Mulberry B by 30 July 1944

houses. American Army engineers were called upon to build five large square-ended caissons at Grays and Tilbury.

During this period at Arromanches, Hickling and Gilbert, by now in a headquarters ashore, were in charge of operations within the port and supervised improvements to the breakwaters, work on which continued until September. By then the port enclosed about two square miles of water and was able to provide berths for seven Liberty ships and twenty-three coasters. The stores pier, able to take a continuous flow of traffic, was linked to twelve pierheads. At the barge pier, less than half a mile long and, according to some staff officers, worth its weight in gold, loads which were too heavy to be carried by DUKWs came ashore usually with the assistance of the floating crane.

Discharge of supplies over the stores pier reached peak at the end of July when the Allies were poised to break out south of Caen. During the last week of that month 136,164 tons of stores came through the Mulberry. They were unloaded by ships' derricks into lorries on the pierhead, or they were discharged by net from derricks into lighters, the loads then being handled by caterpillar cranes on the pierhead and placed on lorries. There is little doubt that with more foresight a higher rate of discharge could have been achieved. Churchill in his original minute about pierheads stipulated that means of accelerating methods of discharge should be considered. This was never done. Moreover, the port operating companies (stevedores from the port of London) were accustomed to working with electrically-operated cranes. In the beginning the loads were small and compact, petrol tins, ammunition boxes and rations; a simple conveyor belt system operated by men used to handling and stacking could have transformed what, in effect, was a slow-moving sequence. Further delay was caused by lack of ship-to-pierhead communications. Eventually a radio link combined with loudhailers to call up ships was devised.

The LST pierhead had only one roadway and was joined to the two-pierhead arrangement. A third pierhead was to have been placed at the seaward end to provide extra space and to take the place of the Baker dolphins which, because of towing difficulties, failed to arrive. According to Witcomb, the skippers of the LSTs (mostly American) some of whom had been more used to driving tractors a year or so before, were enthusiastic about using the pierheads. 'They hated beaching their craft', he writes, 'and were always concerned with damaging their bottoms as the rubbish on the beaches became very great rather quickly.'[4]

By the end of the second week in August, well over 10,000 vehicles (5.5 per cent of the total discharge of British Army vehicles) had passed over the pier. General Sir Harold Pyman, then Brigadier and Chief of Staff to 30 Corps, has recorded that the army landed with little more than 50 per cent of its established vehicles which 'was enough for the job in hand. To wait too long for supplies to build up gives the enemy a chance to build up his. In this event we were just about right and the enemy fell back as we broke out of the bridgehead at full fighting strength.'[5] Apart from the vehicles, some 120,000 troops were brought ashore from infantry landing ships.

But however useful the pierheads were, the DUKWs were more versatile and less labour intensive. There was no bottleneck with them and they had the advantage of unloading where there was plenty of space ashore. From D-Day until 19 November they carried 290,000 tons direct from ship to shore compared with 239,000 tons discharged over the pierheads. In doing so, they wore out their engines and most of them had to be withdrawn early in September. They were, of course, greatly helped by being able to operate in sheltered water.

Omaha had by now been strengthened with a further ten blockships and twenty-one caissons. But a gale on 6 October broke the backs of four of them and twelve Phoenixes were destroyed. In spite of the bad weather, according to the Commander, US Ports and Bases, the Gooseberry continued to be 'of tremendous value.'[6] Ships had to be beached now that the piers had gone but it was discovered that the rock inshore was friable, almost to the consistency of sandstone. Bulldozers planed down these rugged surfaces thus enabling ships to lie ashore without endangering their bottoms on the ebb tide.

Utah also continued to discharge quantities of men, vehicles and supplies. On the last two days of July a record of 25,853 tons were unloaded at the two beaches.[7] (The target tonnages for the two beaches had recently been raised from 5,700 to 10,000 tons for Utah and from 10,500 to 15,000 tons for Omaha.) The flow of vehicles during that month continued unabated. An average of 3,283 per day were brought ashore but as the autumn advanced the tonnages began to decrease. By 1 October it seemed unlikely that the rate of discharge would recover momentum. The weather was becoming increasingly difficult and the DUKWs, most of which had been operating since the early days of the landings, were showing signs of wear and tear. In the British Sector they were moved eastwards after August. Inland behind the beaches rain turned the roads into quagmires. Already the problem of mud had given trouble and now it reached serious proportions. 'Clay accumulated on metalled roads, causing drivers to spin their wheels and puncture tyres as they broke through to the hard, rough foundations.'[8]

By Christmas 1944 dismantling of the Arromanches Mulberry had begun. First to be towed away were the spud pontoons; the piers were dismantled section by section. Many of the girders were at once incorporated into French bridges destroyed by Allied bombing, thus enabling Bailey bridge equipment to be used in the final stages of the campaign. Some of the Phoenixes were raised and towed to Walcheren where they filled gaps in the dykes breached by the heavy bombers of Bomber Command during the subjection of the island. The remainder, together with the blockships, were left to be pounded by the sea. Yet such was the strength of the Phoenixes despite their rapid construction that a number of them can still (1976) be seen from the cliffs above Arromanches*—'grey marine shapes . . . like battlements of a now ruined castle [implying] a violent, bloody history.'[9]

---

*It has been suggested (1976) that they would provide useful data on the effect of sea water on concrete structures now being built for North Sea oil and gas platforms.

10 Contribution to Victory

The building of the Mulberries was probably the greatest military engineering achievement since the Persian crossing of the Dardanelles by a bridge of boats in 480 BC. But the extent to which it helped the victory in north-west Europe is open to question. For an answer, it is worthwhile reviewing, firstly, the construction phase; secondly, the execution of the scheme; and, finally, to attempt to answer the question 'Was it all worthwhile?'

The first phase may be summed up as being a race against time. Its completion was dictated by a shortage of skilled labour and materials like steel. Under such conditions it was scarcely surprising that the components did not arrive on time and thereby prevented the Mulberry forces from training with their equipment. Criticism on this score came mainly from the Americans for they had no responsibility for the construction programme and did not even exist as a force in their own right until the start of the operation. They had, moreover, arbitarily been fitted into a bewildering chain of command, of which the links were chafing against each other. Inevitably they were more sensitive than the British to delays and bottlenecks, nor were they easily assuaged by diplomatic British staff officers whose optimism concealed their own underlying doubts. The delays in the completion of the Phoenix and Whale programmes were attributed to the inability of the British to cope with the programme. The latter's failure to provide adequate equipment to raise the parked Phoenixes was the last straw.

Were the delays unavoidable? Those responsible for the design of the caissons were criticised for too many changes in the design. Ironically, this may have been due to the emphasis on simplicity in the interest of speedy construction.[1] It was subsequently observed that 'many contractors later asked permission to do the very things which had been omitted on this account'. Hence omission of details like splays in the internal walls and stiff horizontal beams at the top of the walls made the caissons more vulnerable to the storms that assailed them.

Whether the Whale equipment, particularly the pierheads,

136

could have been completed more quickly is doubtful. Delay was imposed by last-minute requirements for anti-aircraft defence. The engineering and metal-working industries were subjected to great pressure, particularly due to the shortage of welders and riveters and much preparatory work on the construction sites had to be done. In the end programmes could be completed only by ruthless application of priorities. An engineer later put the matter in perspective. 'The whole of the harbour work including preliminary investigation, design, fabrication and execution of the site, was completed in no more time than would be given normally to the preliminary investigations only for a harbour of the magnitude of that at Arromanches.'[2]

It has been argued that the Mulberry construction programme made substantial inroads on war production at a period of great stress. In fact, the building of the caissons took place in a lull when there were no compelling demands on the building industry; only certain grades of skilled labour were in short supply. The use of dry docks for building caissons, on the other hand, did interfere with the repair and maintenance of shipping and war production suffered from the heavy demands on steel required for the Whale programme.[3] Such items included gun carriages, tanks, jerricans, steel boilers, ammunition boxes and Bailey bridges. The construction of the pierheads, like the caissons, occupied berths which might have been used for the repair of merchant ships. But no interference with the construction of landing craft was permitted.

Before commenting on the execution of the plan, attention must be drawn to the failure to lay down precise responsibilities for the operation of the harbours early enough in the planning stage. The War Office which, in the hands of Tn5, admirably handled the development and construction phases retained this function for too long. And the users of the harbours stated their requirements at too late a stage in the proceedings. Similarly, the navy, which should have had the last word in harbour layout, was consulted too late. A combined naval and military headquarters would have been more effective than Rear Admiral Mulberry/Pluto. (An Allied port committee was required where needs and priorities could have been discussed.)

It would have prevented the fiasco of pumping out the Phoenixes, while knowing exactly what port requirements were necessary would have made siting the breakwaters and general layout much easier.

How effective were the fixed and floating breakwaters? The consensus of opinion was that had there been more time the Phoenixes could have been roofed over and, with small modifications, placed further out to sea, thereby dispensing with the need for Bombardons. As it was, despite their rapid construction, the Phoenixes stood up remarkably well to scour in Mulberry B and did not begin to shift from their locations until the autumn.

The case for the floating breakwaters depended on the provision of berths for deep-draught vessels. But they made heavy demands on steel which could otherwise have been used for constructing floats; they required moorings needed by the caissons to eliminate their grounding off the south coast. Engineers regarded the Bombardons as inefficient at breaking up waves and this was confirmed by post-war experiments. On-the-spot-survey ensured that the Mulberries were large enough to accommodate not only an adequate number of deep-draught vessels but also shelter in bad weather for coastal forces and minesweepers, and they were also used as a ship repair depôt.

The Gooseberries more than came up to expectation, made the least demands on war production and made no demands at all on the hard-pressed tugs. At Utah and Omaha they were used for discharging supplies as late as December 1944 and about six weeks after the British had abandoned the Gooseberries at Courseulles and Ouistreham. An average of 2,000 tons per day passed through Utah and about 15,000 tons per day through Omaha.[4]

The most successful British Gooseberry was at Courseulles where such good use was made of the natural features in siting the blockships. An average of 1,028 tons per day was discharged through this Gooseberry and it was also used for evacuating casualties to ships lying four to five miles offshore.[5]

Turning next to the Whale equipment, about 40 per cent of the pier sections were lost on the cross-Channel voyage.

Fortunately, there were ample reserves of floats and spans, so that the losses did not interfere with erection on the far shore. (It also answers the question that has been asked: whether it would have been possible to complete Mulberry B without the salvaged parts from Mulberry A.) Much of the trouble was caused by flooding of the concrete floats. Although they had to weather much heavier seas than anticipated, some of the damage, at least, was caused by 'reckless and brutal' handling and this must be attributed to the lack of time for training. (Inadequate training may have been also responsible for careless assembly of spans in the latter stages.) There was also insufficient timber for protecting the floats.[6]

The stores pier was criticised on the grounds that it was too elaborate a structure for its function. At Arromanches only about 15 per cent of stores was discharged from coasters into lorries on the stores pier. Ships with draughts greater than 24ft could not berth alongside it. While urgent stores, mail and certain awkward or heavy loads were discharged over this pier, the Monckton report[7] (which attempted to assess the value of Mulberry) concluded that the LST pier was more useful in spite of its relatively short life and should have been given higher priority. By the end of October 39,000 vehicles had passed over it and 220,000 soldiers had landed dryshod. An LST could be cleared in about 23 minutes as compared with $2\frac{1}{2}$ hours if it had been beached; in the latter case it had to wait 6 hours to be refloated by the next high tide. This was apparently faster than the Americans who took about 40 minutes longer. With higher priority for the LST pier it would, together with the barge pier, have been sufficient. Urgent stores could have been brought ashore by DUKW. This, however, did not take into account the two-way system on the stores pier which kept traffic flowing. With more sophisticated unloading techniques (such as conveyor belt systems) the flow would have been still faster. The Americans in the two-day life of their LST pier made good use of it, though they too had no special unloading equipment at that time.

The pierheads served their purpose as a rapidly devised piece of military equipment and there appeared at the time to be no further justification for their survival. However, they have proved to be the forerunners for the jack-up equipment used in oil-drilling operations in recent years. The buffer pontoons were admirable for discharge of the lower decks of LSTs, though at the British pierheads caution prevailed (there was less urgency than in Mulberry A) and instead of driving their vessels at three knots up the ramp and thus possibly endangering the pierhead, skippers were content to be warped into position.

Finally, how effective was Mulberry? Discharge figures for Mulberry B show that by the end of October 25 per cent of stores, 20 per cent of personnel and 15 per cent of vehicles were landed through the artificial harbour.[8] The remainder came ashore over open beaches (protected by the Gooseberries) and through Port en Bessin (captured intact) and Ouistreham, and later through ports in the Pas de Calais. Given the quantity of landing craft and DUKWs that ultimately became available, the pier, though not the blockships or the caissons, could have been omitted, as was proved at Utah and Omaha.

But, bearing in mind that in the planning of Overlord the gaining of surprise was paramount, Allied strategy depended on landing on open beaches. Artificial harbours (costing around £25 million) were an insurance without which, as Maj-Gen Sir Frederick de Guingand, Chief of Staff to Montgomery, declared, it would have been wrong to have undertaken the operation.[9] Once this had been established, there had to be sheltered water provided not only by blockships but by caissons in deeper water. And without large numbers of amphibious craft (as was the case in the early stages of planning) it would have been wrong not to have provided means of discharging men, vehicles and stores.

Moreover, the failure of Montgomery to turn the German flank at Arnhem and the delay in opening Antwerp, demonstrated the value of Mulberry in the late summer and autumn of 1944. It was later suggested by a civilian, Bosworth Monck, a member of the Joint Planning Staff and the Ministry of Supply, that the problem might have been solved by a more flexible use of the breakwaters.[10] By refloating the caissons off Normandy they could have accompanied the advance eastwards up the coast, like a siege train, and have been used, for instance, at Ostend, to provide sheltered water for the discharge of supplies from coasters. Thus an outflanking operation might have proved easier and the war in Europe shortened. (The lack of tugs, to mention only one factor, would have made this operation impossible.)

The senior commanders were at least grateful for the Mulberries. According to Eisenhower, 'Mulberry exceeded our best hopes ... average tonnage per day from 20 June to 1 September was 6,765 tons.'[11] Eisenhower's deputy, Air Chief Marshal Sir Arthur Tedder, acknowledged Churchill's persistence in advocating the Mulberries without which 'that extraordinary operation might never have come to fruition. Indeed the whole question of the invasion of Europe might well have turned on the practicability of these artificial harbours.'[12] And although there is force in the detractors of Mulberry's arguments, they ignored, as Australian war correspondent and military historian, Chester Wilmot, has ably argued, important strategic and psychological factors.[13] 'Strategically the "possession" of Mulberry gave the planners the freedom to choose a landing area well away from the heavily fortified major ports; psychologically, it gave the Allied High Command a degree of confidence without which the venture, which seemed so hazardous, might never have been undertaken.' Albert Speer, Reichsminister for Armament and War Production, supported this argument when he concluded that having by-passed the Atlantic Wall 'by means of a single brilliant technical device', the Allies made the German defence system completely irrelevant.[14]

In conclusion, it is worth remembering that, as Col (later Maj-Gen) L. D. Grand, RE, one of the founders of the Special Operations Executive, and after the war Director of Fortifications and Works, put it, almost impossible problems in Mulberry were solved 'not by soldiers or by civilians but by British engineering.'[15]

Select Bibliography

Brownlowe, L. C. *The Mulberry Project* (Badger Books.) The only work of fiction, so far as is known, devoted to Mulberry.

Bykofsky, J. and Larson, H. *The US Army in World War 2. Transportation Corps Operations Overseas* (Washington, 1957).

Churchill, Winston S. *The Second World War*, vol 5 'Closing the Ring', p 195 (1952).

Instn of Civil Engineers. *The Civil Engineer in War*, vol 2, Docks and Harbours, 1948. A collection of papers read at a symposium by engineers involved in the development of the Mulberry components. The discussions are of particular value.

Cornish, Henry. *Dock and Harbour Engineering*, vol 2. Design of Harbours (1959).

Cowan, R. J. P. 'Notes on the Construction of Mulberry Harbour in Normandy, June–July 1944'. Lecture Imp Coll of Science & Technology (June 1975).

Eisenhower Foundation. *D-Day. The Normandy Invasion in Retrospect* (University Press of Kansas, 1971).

Ellis, L. F. *et al. Victory in the West*, vol 1. (Official History of the Second World War, Military Series) (1962).

Ellsberg, Edward. *The Far Shore* (1961). Valuable as an account by an experienced American salvage expert on aspects of the Mulberries.

Fergusson, Bernard. *The Watery Maze*. The story of Combined Operations (1961).

Harrison, Gordon A. *Cross-Channel Assault* (Washington, 1951). Official History of the US Army in the Normandy landings.

Harrison, Michael. *Mulberry. The Return in Triumph* (1965). A discursive account largely from the Tn5 standpoint, but unaccountably stopping at D-Day. Little detail on equipment.

Hartcup, Guy. *The Challenge of War*. Scientific and Engineering Contributions to World War Two (1970).

Hickling, Harold. *Sailor at Sea* (1965). References to the Mulberries more fully developed in *Story of the Mulberries*, Harold Hickling and Ian Mackillop, produced by the War Office in 1945 and now in the PRO.

Hickling, H. 'The Prefabricated Harbour', *Jnl Roy United Services Instn* (Aug 1945).

Hodge, W. J. 'The Mulberry Invasion Harbours'. *The Structural Engineer*, Jnl of the Instn of Structural Engineers (March 1946). A good account of the construction of the Phoenixes.

Inman, P. *Labour in the Munition Industries.* (Official History of the Second World War, Civil Series) (1952).

Kohan, C. M. *Works and Buildings* (Official History of the Second World War, Civil Series) (1952).

Mallory, Keith and Ottar, Arvid. *Architecture of Aggression* (1973). An interesting study of military building projects in both world wars. Contains a section on Mulberry.

McGivern, Cecil. *The Harbour called Mulberry.* The script of a BBC feature programme. (Pendulum Publns, 1945.) An imaginative and vivid, yet accurate account.

Parker, H. M. D. *Manpower* (Official History of the Second World War, Civil Series) (1957).

Pawle, Gerald. *The Secret War.* The story of the Department of Miscellaneous Weapons Development. It traces the story of Bubble, Lilo and Bombardon (1956).

Postan, M. M. *British War Production* (Official History of the Second World War, Civil Series) (1952).

Roskill, Stephen. *The War at Sea*, vol 3, Pt I (Official History of the Second World War, Military Series) (1960).

Ruppenthal, Roland G. *The US Army in World War 2. The European Theatre of Operations. Logistical Support of the Armies*, vol 1 (1953); vol 2 (1959). The only official Allied military history which deals exclusively with the logistical problems of the north-west European campaign.

Stanford, Alfred. *Force Mulberry.* The Planning and Installation of the Artificial Harbour off US Normandy Beaches in World War II, (New York, 1951).

Thompson, R. W. *The Price of Victory* (1960).

Wernher, Sir Harold. *World War 2. Personal Experiences.* Privately published (1950). Though brief, contains insights on the friction generated during the construction of the Mulberry components.

Wilmot, Chester. *The Struggle for Europe* (1952).

References

Key to Abbreviations

A number of documents on Mulberry are available for study in the Public Record Office (PRO). They are classified under the following headings: Admiralty (ADM); Cabinet (CAB) including papers of the Chiefs of Staff; Joint Planning Staff; Joint Staff Mission in Washington; Defence (DEF). DEF2, in particular, contains the papers of COHQ. The Prime Minister's files (PREM) reflect the particular interest which Winston Churchill took in the development of Mulberry. The Ministry of Supply (SUPP). There was little systematic attempt to preserve records in this largely wartime creation. War Office (WO) contain the papers of Tn5 and Chiefs of Staff papers relating to the development of Mulberry.

CHAPTER ONE

1 White, Sir Bruce G. 'The Mulberry Harbours', *The Central* (Journal of the Old Centralians) No 94, Dec 1946.
2 White, Sir Bruce G. 'Construction and Operation of Military Ports in Gareloch and Loch Ryan', *Instn Engrs & Shipbuilders Scotland Procs*, Jan 1949.
3 Pawle, Gerald. *The Secret War*, passim, 1956.
4 Fergusson, Bernard. *The Watery Maze*, pp 285–6, 1961.
5 Skerrett, R. G. 'The Navy's Steel Pontoons', *Compressed Air Mag*, Sept 1945; PRO/DEF2/1063.
6 PRO/CAB88/15; PRO/PREM3/216/7.
7 PRO/CAB88/15; PRO/DEF2/144.
8 Davies, John Langdon. *Sir H. Wernher Bt, GCVO*, chap 7 (privately published), 1948.
9 PRO/WO32/12211.
10 PRO/WO32/12211; Wernher, Sir H., *World War 2. Personal Experiences* (privately published), 1950.
11 PRO/WO32/12211; PRO/ADM199/1614; White, Sir Bruce. Disc on breakwaters, *The Civil Engineer in War*, 1948, vol 2, pp 316–17.
12 Seabees in Normandy Invasion. *Eng News Rec*, 28 Dec 1944, pp 823–30. US Navy Department. 'Building the Navy's bases in World War 2, 1940–46', vol 2, pp 99–117, 1947.

CHAPTER TWO

1 PRO/PREM3/216/1.
2 Ibid; PRO/DEF2/438; PRO/DEF2/58.
3 PRO/DEF2/907.

4   PRO/DEF2/948–950; Hamilton, R. M. 'Floating Wharves and Jetties', *Dock & Harbour Authy Jnl*, April 1946; Pawle, op cit, chaps 18–19.

5   Beckett, A. H. 'Some Aspects of the Design of Flexible Bridging, including Whale Floating Roadways', *The Civil Engineer in War*, vol 2, op cit, pp 385–400, 1948.

6   Bruce White & Beckett—Everall correspondence. April–June 1943 (Beckett Coll).

7   Wood, C. R. J. 'Reinforced Concrete Pier Pontoons & Intermediate Pierhead Pontoons', *The Civil Engineer in War*, Vol 2, op cit, pp 401–16, 1948; correspondence L. G. Mouchel & Partners—Everall (Mouchel Coll).

8   Findlay's of Motherwell. 'Mulberry Pierheads. How They Were Built', (undated brochure).

9   Pavry, R. 'Mulberry Pierheads', *The Civil Engineer in War*, vol 2, op cit, pp 371–3.

10 PRO/PREM3/216/1.

11 Beckett, op cit, pp 396–8; War Office publn. 'Notes on Floating Bridge Equipment', February 1944.

12 Progress Repts, Cairnhead; PRO/WO166/12073.

13 Bruce White & Beckett—Everall correspondence, op cit.

14 Pavry, R., op cit, p 375.

15 Pavry, R., op cit, pp 380–2; Baker, A. L. L. 'The Heysham Jetty', *Instn Civil Engrs, Session 1947–48*; Baker, A. L. L. 'The Heysham Jetty Suspended Fenders', *Concrete & Constructional Engng*, June 1946; Gravity Fenders, Pat No 563946, 6 Sept 1944.

16 Baker, A. L. L. 'Folding Dolphin', Pat No 573027, 2 Nov 1945.

17 PRO/DEF2/1063.

CHAPTER THREE

1   PRO/DEF2/529.

2   PRO/DEF2/438; PRO/DEF2/945.

3   PRO/DEF2/144.

4   Lochner, R. A. Typescript (Mary Lochner Coll); Pawle, op cit, pp 240–53.

5   PRO/CAB119/83.

6   Misc Weapons Dept Rept MWD/E78/1.

7   Churchill, Winston S. *The Second World War*, vol 2. 'Their Finest Hour', 1949, chap 10, pp 201–3; Posford, J. A. 'Construction of Britain's Sea Forts', *The Civil Engineer in War*, vol 3, op cit, p 132 et seq 1948.

8   Gwyther Papers, (Coode & Pturs Coll).

9   PRO/WO32/12211.

10 Wilson, W. Storey. Ms 16 Jan 1975; 'Monolith Construction', Prodn Cttee Rept, 4 Oct 1943 (W. S. Wilson Coll).

11 Gwyther Papers, op cit.
12 Hodge, W. J. 'The Mulberry Invasion Harbours. Their Design, Preparation and Installation', *The Structural Engineer*, March 1946, p 146 et seq.
13 Todd, F. H. 'Model Experiments on Different Designs of Breakwaters', *The Civil Engineer in War*, vol 2, op cit. 'Some Model Experiments carried out in connection with the Mulberry Harbour', *Trans Roy Inst Naval Arch*, 1946, p 196.
14 Hodge, W. J., op cit.
15 Gwyther Papers, op cit.
16 Little, D. H. 'Mulberries', *Navy Works* (Jnl of Civil Engineer-in-Chief's Dept, Adm), Aug 1949, pp 22–9.

CHAPTER FOUR

1 Adm Memo. 'Artificial Harbours in Opn Overlord', 6 March 1945.
2 Gwyther Papers, op cit.
3 PRO/PREM3/216/7.
4 Hickling, H. & Mackillop, I. *Story of the Mulberries* (WO), chaps 5 and 9, 1945.

CHAPTER FIVE

1 Postan, M. M. *British War Production*, 1952, pp 280–4; Kohan, C. M. *Works and Buildings*, 1952, p 246 et seq; Parker, H. M. D. *Manpower*, 1957, chap 14.
2 Wigmore, V. S. Correspondence with author, Jan 1976; Presidential Address to Soc of Engrs, 1947.
3 Stork, S. G. & Cowan, T. O. 'Nos 1 and 2 Port and Inland Water Transport Repair Depots', *The Civil Engineer in War*, op cit, vol 2, pp 100–11, 1948.
4 Mitchell, C. D. 'Prefabricated Precast Concrete Structures including Floating Craft', *Trans Soc of Engrs*, Oct 1947.
5 Wernher, Sir H. *World War Two: Personal Experiences* (privately published), 1950.
6 War Office Publn. 'Notes on Floating Bridge Equipment', 1944, op cit.
7 Stork & Cowan, op cit.
8 Moon, A. R. 'Welding during the War and After', *Structural Engineer*, June 1947.
9 Findlay's brochure, op cit.
10 Author's interview with Dr O. Kerensky, 4 March 1967; *North Wales Wkly News*, 17 May 1945.
11 Witcomb, E. W. to author, 30 Dec 1975.
12 PRO/DEF2/501; *Shipbldg & Shipping Rec*, 15 Feb—10 May 1945.
13 PRO/WO165/106 Pt III.

14   Stanford, A. to author, 24 Aug 1975.
15   Stanford, A. *Force Mulberry*, New York 1951, pp 110–12.
16   Pavry, op cit. p 379; Pavry to author, 22 July 1976.
17   Mitchell, C. D., op cit.
18   Wigmore, V. S. to author, 26 Jan 1976.
19   Hodge, W. J., op cit, p 191.
20   Hawkey, R. G. 'E. India Import Dock and S. Dock Surrey Comm Dock', *Civil Engineer in War*, vol 2, op cit, pp 36–50, 1948; Fuller, W. V. 'Mulberry', *Jun Instn of Engrs*, April 1946.
21   Production Committee Rept, op cit, 4 Oct 1943.
22   Gwyther Papers, op cit.
23   PRO/PREM3/216/7.
24   Fuller, W. V., op cit.
25   Moon, A. R., op cit.
26   Misc Weapons Dept Rept, MWD/E78/1, op cit.
27   Beckett, A. H. to author, 12 July 1976.

CHAPTER SIX

1   Gwyther Papers, op cit.
2   PRO/CAB88/15.
3   Stanford, A. to author, 20 July 1975.
4   Stanford, A. *Force Mulberry*, op cit, p 75, 1951.
5   PRO/DEF2/428.
6   Stanford, A. to author, 20 July 1975.
7   Gwyther Papers, op cit.
8   Tn5. Phoenix. General Description and Notes on Sinking and Raising, 6 April 1944.
9   Rolfe, J. A. S. to author, 2 June 1965.
10   PRO/ADM199/1618.
11   Lipscomb, F. W. and Davies, John. *Up She Rises. The Story of Naval Salvage*, 1966, chap 5.
12   Ellsberg, Rear Adm E. *The Far Shore*, 1961, chaps 1–13.
13   PRO/ADM199/1618.
14   Ibid.
15   Ellsberg, op cit, p 55.
16   Misc Weapons Dept Reprt, op cit.
17   PRO/ADM199/1617.

CHAPTER SEVEN

1   PRO/ADM53/119240.
2   PRO/ADM199/1615.
3   Cowan, R. J. P. 'Notes on the Construction of Mulberry Harbour'. Lecture Imp Coll of Science & Technology, Dept of Civil Engng, 17 June 1975.
4   Harrison, G. A. *Cross Channel Assault*, OCMH, Washington, 1951 passim.

5   Mais, Rt Hon Lord. Notes for author, Aug 1975.

6   Hickling & Mackillop, op cit, chap 5.

7   Adm Memo. 'Artificial Harbours in Opn Overlord', 6 March 1945, sect III.

8   Jellett, J. H. 'The Layout, Assembly and Behaviour of the Breakwaters at Arromanches Harbour (Mulberry B), *The Civil Engineer in War*, vol 2, op cit, pp 291–312, 1948.

9   Adm Memo. 'Artificial Harbours in Opn Overlord', op cit, Sect II; Rept to Controller Adm from DMWD, 16 June 1944.

10   Ruppenthal, R. G. *The US Army in World War 2, European Theatre of Operations*. Logistical Support of the Armies, vol 1, 1953, chap 10.

11   Jellett, J. H., op cit.

12   PRO/WO171/1754; Cowan, R. J. P., op cit; Sainsbury, J. R. to author, 29 July 1976.

13   PRO/DEF2/428.

14   PRO/WO171/1733.

15   PRO/WO171/1737.

16   Stanford, A. to author, 20 July 1975.

17   Mais, Rt Hon Lord. Notes, op cit; Beckett, A. H. to author, 12 July 1976.

18   Ellsberg, E., op cit, pp 304–8.

19   Stanford, A. *Force Mulberry*, op cit, pp 174–5, 1951.

20   Baxter, J. P. *Scientists against Time*, 1968 edn MIT Press paperback, p 79; *The Story of the RASC, 1939–45*, 1955, pp 598–602.

21   PRO/ADM53/119240; Gilbert, S. K. Notes for author, Oct–Nov 1975.

22   Hickling & Mackillop, op cit, chap 13, 1945.

23   PRO/DEF2/501 (SHAEF Handbook Mulberry B, 5 Nov 1944); Ruppenthal, R. G., op cit, vol 1, chap 10.

CHAPTER EIGHT

1   PRO/DEF2/428.

2   Stagg, J. M. *Forecast for Overlord*, 1971, pp 125–6.

3   Walter, Brig A. E. M. to author, 25 Oct 1975.

4   Mais, Rt Hon Lord. Notes for author, op cit.

5   Cowan, R. J. P., op cit.

6   Pakenham-Walsh, Maj Gen R. P. *History of the Corps of Royal Engineers*, vol 9, 1938–48, 1958.

7   Stanford, A. *Force Mulberry*, op cit.

8   Ruppenthal, R. G., vol 1, op cit, pp 406–13, 1953.

9   Ibid. p. 413; Harrison, G. A., op cit, pp 423–6.

10   Mais, Rt Hon Lord. Notes for author, op cit.

11   PRO/WO171/1418.

CHAPTER NINE

1 Supreme Commander's Rept to Comb Chiefs of Staff on Operations of the AEF in Europe, 1946, p 75.
2 Ruppenthal, R. G., vol 2, op cit, pp 53–62, 1959.
3 Hickling, Capt H., RN. Rept on Mulberry B, 19 Aug 1944; Jellett, J. H. Rept to Civil Engineer-in-Chief (Imp War Mus Coll–HH3).
4 Witcomb, E. W. to author, 30 Dec 1975.
5 Pyman, Gen Sir H. *Call to Arms*, 1971, p 74.
6 Adm Memo. 'Artificial Harbours in Opn Overlord', op cit, 1945. App 3. Cdg Off US Naval Base 11 to Cdr US Ports and Bases, 1945, France.
7 Ruppenthal, R. G., op cit, vol 2, pp 54–5.
8 Ibid., vol 2, p 57.
9 Powell, Anthony. *The Military Philosophers*, 1969, p 169.

CHAPTER TEN

1 Hodge, W. J., op cit, p 150.
2 Nowlan, H. J. Disc on Mulberry Breakwaters, *The Civil Engineer in War*, vol 2, op cit, pp 326–7, 1948.
3 Postan, M. M. *British War Production*, op cit, pp 280–4, 1952.
4 Ruppenthal, R. G., op cit, vol 2, p 57.
5 Adm Memo. 'Artificial Harbours in Opn Overlord', op cit, Sect III, 1945.
6 Wood, C. R. J. to author, 23 March 1974.
7 PRO/ADM199/1616.
8 Pakenham-Walsh, R. P., op cit.
9 PRO/DEF2/501.
10 Ibid.
11 Supreme Cdr's Rept, op cit, p. 69.
12 Tedder, The Lord, *With Prejudice*, 1966, p 13.
13 Wilmot, Chester, *The Struggle for Europe*, 1952, pp 387–8.
14 Speer, Albert. *Inside the Third Reich*, 1970, pp 352–3.
15 Grand, Col L. D. Disc on Mulberry Components, *The Civil Engineer in War*, vol 2, op cit, pp 445–6, 1948.

Appendix

WIND AND SEA CONDITIONS IN THE VICINITY OF THE MULBERRY HARBOURS, JUNE TO JULY 1944

Date	Wind, Direction and Force	Mean Height of Waves
June		*Feet*
6	WNW–NW force 4	2–3
7	NW–NNW force 4 decreasing to force 2–3 pm	2–3 decreasing to 1–2
8	NW force 2–3 backing nearly WSW–SW force 3–4	1–2
9	WSW force 2–3 veering W–NNW force 3–4 pm and decreasing to 2–3 after about 20.00hrs	2–3
10	WNW–NW force 3	2
11	NW force 2–3 falling light variable and becoming SW–W force 3–4 pm	0–2
12	W force 3–4 backing SW force 2–3	1–3
13	S–SW force 2–3 freshening to force 4–5 and veering W pm	0–2 increasing to 3–4
14	WSW–WNW force 4–5 moderating to force 4–5 after about 18.00hrs	3–4 decreasing to 2–3
15	W–WNW force 3–4 backing WSW–SW pm	2–3
16	WSW–SW force 3–4 increasing to force 4–5 and veering WNW–NW about 15.00hrs and moderating to force 3–4	2–3 increasing to 3–4
17	NW–N force 3–4 reaching force 4–5 at times	3–4
18	NNW force 3–4 veering N–NE force 2–3	2–3 decreasing to 1–2
19	N–NE force 3–4 freshening to force 5–6 by about noon and reaching force 6–7 at times	2 increasing to 5–6
20	NE force 5–6 freshening to force 6–7	5–6 increasing to 8–9
21	NE force 6–7 moderating slowly after about 09.00hrs to force 5–6	8–9 decreasing slowly to 2–3
June		*Feet*
22	NE force 5–6 moderating gradually to force 3–4 by about 20.00hrs	5–6 decreasing slowly to 2–3
23	N–NE force 2–3	2–3

24	N–NE force 2–3 falling light variable	2 becoming 0
25	Light variable at first becoming S–SW force 2–3 and SW–W force 3–4 pm	2–3 increasing to 3–4
26	SW force 2–3 backing S–SW force 3–4 from about 11.00hrs and reaching force 4–5 during afternoon	2–3 reaching 3–4 pm
27	SW–W force 3–4 freshening to force 4–5 pm	2–3 increasing to 3–4
28	SW force 4–5 increasing to force 5–6 between 01.00 and 18.00hrs and moderating to force 3–4 late pm	4–5 increasing to 6–8 and later decreasing to 4–5
29	S–SW force 3–4 decreasing to force 2–3	4–5 decreasing to 1–2
30	SW–WSW force 3–4	2–3

July

1	S–SW force 3–4 occasionally reaching force 5 early	2–4
2	S–SW force 3–4	2–3
3	S–SW force 2–3 veering W–NW force 3–4 am	1–2 increasing to 2–3
4	W–NW force 3–4 backing SW–W force 3	1–2
5	S–SW force 3–4 backing S force 2–3 pm	1–2
6	S–SE force 2–3 freshening to force 4–5 from about 12.00 to 22.00hrs	1–2 increasing to 3–4
7	S–SE force 2–3 freshening to force 4–5 after about 16.00hrs	2–3 increasing to 3–4
8	S force 3–4 falling light variable after about 16.00hrs	3–4 decreasing to 0–2
9	Light variable at first becoming SW–W force 4–5 by about noon	0–1 increasing to 3–4 pm
10	W force 4–5 veering W–WNW force 5–6	3–4 increasing to 5–6
11	W–NW force 5–6 moderating slowly to force 4	5–6 decreasing to 4–5
12	W–NW force 2–3 backing W–SW force 3–4	3–4 decreasing to 2–3
13	W force 2–3 increasing to force 3–4	2–3
14	W force 3–4	2–3
15	W–SW force 2–3 at first increasing to force 3–4	1–3
16	W–SW force 3–4 decreasing to force 2–3 by about 12.00hrs and becoming light variable by pm	1–3 decreasing to 0–1

17	Light variable becoming E force 3–4 pm	0–1 becoming 1–2
18	E force 3–4 falling light variable late pm	2–3 decreasing to 1–2
19	Light variable	0–1
20	Light variable early becoming NE–E force 2–3	0–1 increasing to 1–2
21	NE–E force 2–3 increasing to force 3–4 and reaching force 5 at times	1–2 increasing to 3–4
22	N–NE force 3–4 backing slowly NW–N and decreasing to force 2–3	3–4 decreasing to 1–2
23	N–NW force 2–3 falling light variable	1–2 decreasing to 0–1
24	Light variable becoming mainly southerly force 2–3 pm	0–1
25	SE force 2–3 freshening to force 3–4	0–1 increasing to 2–3
26	S force 3–4 veering SW force 3–4 am	
27	SW–W force 3–4	No observations made but mean height of waves unlikely to have exceeded 3ft at any time
28	W force 3–4 backing slow by SW force 3–4	
29	S–SW force 3–4 veering SW–W force 3–4	
30	SW–W force 3–4 veering W–NW force 3–4 and becoming light variable late pm	
31	Light variable becoming S force 2–3 pm	

NOTE. *The maximum height of individual waves can be taken as about 40 per cent greater than the mean height given in this column.*

Acknowledgements

My thanks are due, firstly, to those who were associated with the development and construction of the Mulberry components, or who installed them off the Normandy beachhead, and gave me their valuable help either in the form of recollections, the loan of documents and photographs, or criticism of the first draft of the book. They include A. R. W. Adcock, Professor A. L. Baker, A. H. Beckett, W. H. Booker, the late J. G. Carline, J. Crawford, R. J. P. Cowan, T. O. Cowan, S. W. Cox, Col S. G. Gilbert, J. R. Gwyther (son of R. G. Gwyther) and, through him, Coode & Partners (Consulting Engineers) who hold the papers of the late R. G. Gwyther, I. Hughes, Dr Oleg Kerensky, the late D. H. Little, Mrs M. Lochner, the Rt Hon Lord Mais, E. R. Mills, C. D. Mitchell, R. Pavry, J. A. S. Rolfe, J. R. Sainsbury, Capt A. Stanford, USNR (retd), D. J. Tonks, Brigadier A. E. M. Walter, Lady Zia Wernher, Sir Bruce White, V. S. Wigmore, W. S. Wilson, E. W. Witcomb and C. R. J. Wood.

Secondly, I must thank the staffs of the libraries and other institutions holding material on Mulberry harbours. They include the Public Record Office, the invariable courtesy and helpfulness of whose staff make all who research there indebted to them, the Imperial War Museum, the Librarians and staff of the Institution of Civil Engineers, the Institution of Mechanical Engineers, in particular Mr S. G. Morrison and Mr L. T. Griffith, Secretary of the Society of Engineers, for allowing me to quote from the Society's proceedings.

I am further indebted to the following for permission to use illustrations: the Public Records Office for all the figures with the exceptions given below, and for plates I, III, VII, XV and XVI, all of which are Crown Copyright; figures 13, 14 and 19 are reproduced from *The Civil Engineer in War* by kind permission of the Institution of Civil Engineers; the Imperial War Museum for plate XVIII. The rest come from my own and other private sources of friends all of whom I thank once again.

Index